HAZARDOUS

Captain

7825 Harrisburg
Houston, TX 77012
(713) 928-5806

For My Grandchildren…

Richard Thomas Bardsley

Ashley Lyn Ditzel

Kevin Bradley Krivitsky

Kimberly Lynne Krivitsky

Keith Daniel Lord

Derek Ryan Stradley

Justin Royce Stradley

…Who May Never See A Fire Alarm Box

FRONT COVER:
Fire background — Capt. William Noonan, Boston Fire Dept.
Alarm Box — Mike Tonegawa
Design and Layout — Ron Grunder

SELECTED BIBLIOGRAPHY OF PAUL DITZEL BOOKS

A Century of Service — The History of the Los Angeles Fire Department

Emergency Ambulance

Fire Alarm! The Fascinating Story Behind The Red Box On The Corner

Fire Alarm! The Story of a Fire Department

Fire Engines, Firefighters

Fireboats, A Complete History of the Development of Fireboats in America

How They Built Our National Monuments

Railroad Yard

The Complete Book of Fire Engines

The Day Bombay Blew Up

The Kartini Affair

True Blooded Yankee

Library of Congress Cataloging in Publication Data
Ditzel, Paul

Fire Alarm!
Library of Congress Catalog Number: 89-080108
ISBN 0-925165-02-6

Published by Fire Buff House Division of Conway
Enterprises, Inc.
P.O. Box 711, New Albany, Indiana 47151

Printed in the United States of America

FIRE ALARM!

The Fascinating Story Behind The Red Box On The Corner

By Paul Ditzel

A Fire Service History Series Book From Fire Buff House Publishers
Div. of Conway Enterprises, Inc., New Albany, Indiana 47151

CONTENTS

THE GAMEWELL FIRE ALARM TELEGRAPH COMPANY

GENERAL OFFICES AND WORKS, NEWTON UPPER FALLS, MASS.

FOREWORD

Bright red Gamewell fire alarm boxes were on tens of thousands of street corners — in villages, towns, and cities — throughout America for more than a century. With a 95% market share, Gamewell held a virtual monopoly on the nation's fire alarm systems.

Never before has the fascinating story behind the red box on the corner been told. Most of the boxes are gone now, relics of the past; but their story is one of intrigue, ingenuity, and even pathos. One Gamewell principal stayed on the job daily until age 95, while another, whose patents were in a large measure responsible for Gamewell's success, took his own life.

Millions of fire alarms have been turned in from the boxes on America's street corners — alarms for everything from a wisp of smoke quickly snuffed out with a shot of "chemical" from the extinguisher carried on the running board of the first arriving engine, to when Mrs. O'Leary's immortal cow kicked over the lantern in Chicago, to when the San Francisco earthquake struck, or to when entire cities were nearly consumed by flames.

As a small boy, I would often look up at the box on the corner near our house and wonder if it were true that a pair of handcuffs would descend from the box to entrap anyone sending an alarm until the firemen arrived. Many of my little friends believed such was the case and avowed they would never turn in an alarm — even for a real fire. But I knew if I ever saw a fire I would turn in an alarm no matter what happened, and I even kept a rock hidden in the bushes near the box so that I could easily break the glass. I often checked to make sure the rock was still there. Perhaps it is there still, as I haven't checked since I moved away nearly fifty years ago. But a recent visit showed the box is there yet — an ever faithful sentinel on the corner, still in service.

Little did I dream that inside the fire alarm box was an intricate, clock-like mechanism which caused amazing things to happen when the hook was pulled. This book chronicles those amazing things, as well as giving the complete story of fire alarm telegraph from its very beginning.

How delighted I am that the man who is considered to be the "dean" of fire historians, and who is the winner of many awards for his books on fire service subjects, undertook the task of recounting the story of fire alarm telegraph. Paul Ditzel, author of **Fire Engines, Firefighters, A Century Of Service, Fireboats,** and many others, has written a comprehensive and compelling account of fire alarm telegraph. If this book is only half as interesting to you as it has been to me, you will be unable to put it down until the last page is finished.

W. Fred Conway, September 1990

W. Fred Conway, a former Fire Chief, and a lifelong fire buff, is the author of ***Chemical Fire Engines,*** *a type of fire apparatus, which, like fire alarm boxes, were once prevalent throughout America, but which have also faded into the past. Chief Conway maintains an extensive collection of fire alarm telegraph instruments, which are displayed along with antique fire engines and other early fire service memorabilia in his privately owned museum in the Conway Enterprises Building in New Albany, Indiana. The Museum is open by appointment without charge.*

FIRE &
POLICE
TELEGRAPHS,
23
THE
GAMEWELL
Fire Alarm Telegraph
Company,
GENERAL OFFICES, NEW YORK.
19 BARCLAY ST.

Buffalo, New York's infamous bad luck alarm, Hoodoo Fire Alarm Box No. 29, was sounded for this blaze in the Seneca Electric Co. As midnight neared on December 30, 1981, three alarms were struck for the box in accordance with the running card. Over more than a century, alarms from Box 29 have meant destruction, injury, and death — a jinx unique in the American fire service.

CHAPTER ONE

HOODOO FIRE ALARM BOX 29

Chilling winds from off Lake Erie swept Buffalo while snow flurries added to the white blankets splotching the city of 400,000 people. Considering the nasty weather that night, December 30, 1981, firemen hoped alarm bells would remain silent.

Bad weather often means bad fires: overheated furnaces... homeless people kindling warming fires in old firetrap buildings... deserted streets enabling fires to grow big before discovery. There had not been a major fire in Buffalo for five days, but firemen knew their luck could not hold much longer.

Shortly before 11 p.m., dispatchers in the Ellicott Street Fire Alarm Office answered telephone and 9-1-1 system calls reporting a fire at 166 Seneca Street in the conflagration-breeding nest of antiquated buildings in the commercial district just east of downtown. The location was entered into a computer which flashed the numbers of the nearest fire alarm box and apparatus on a video display terminal similar to a television screen.

At 10:52 p.m., alarm tones peeped in firehouses. By loudspeakers and radio a dispatcher broadcast: "Alarm of fire... 166 Seneca Street near Elm... Box 29... The following companies will respond: Engines 1, 13 and 32; Ladders 2 and 1; Rescue 1; the First Battalion Chief and Division Chief 1..."

As firemen scrambled for their turnouts and boots, a dispatcher pushed the Number 2 and the Number 9 buttons on an alarm repeater and pressed the transmit button. Electrical impulses, opening and closing circuits, caused firehouse bells to ring twice, pause, then strike nine more blows.

The same electromechanical system simultaneously spewed thin strips of paper tape from fire station alarm registers. With each stroke of the bell, the clock-like mechanisms whirred as the registers punched two holes in each tape, left a space, then stabbed nine more holes.

When Buffalo began removing its fire alarm boxes, Hoodoo Box 29 was preserved and mounted at the Buffalo Fire Department Museum. Although many Gamewell boxes had preceded it, the last Hoodoo Box was of a 1931 style with an exterior shell of lightweight Herculite and a white-painted quick action door. It was activated by pulling down the handle and then a lever. Paul Ditzel

Ellicott Street Fire Alarm Office of the Buffalo Fire Department, where all alarms are received. Jack Supple

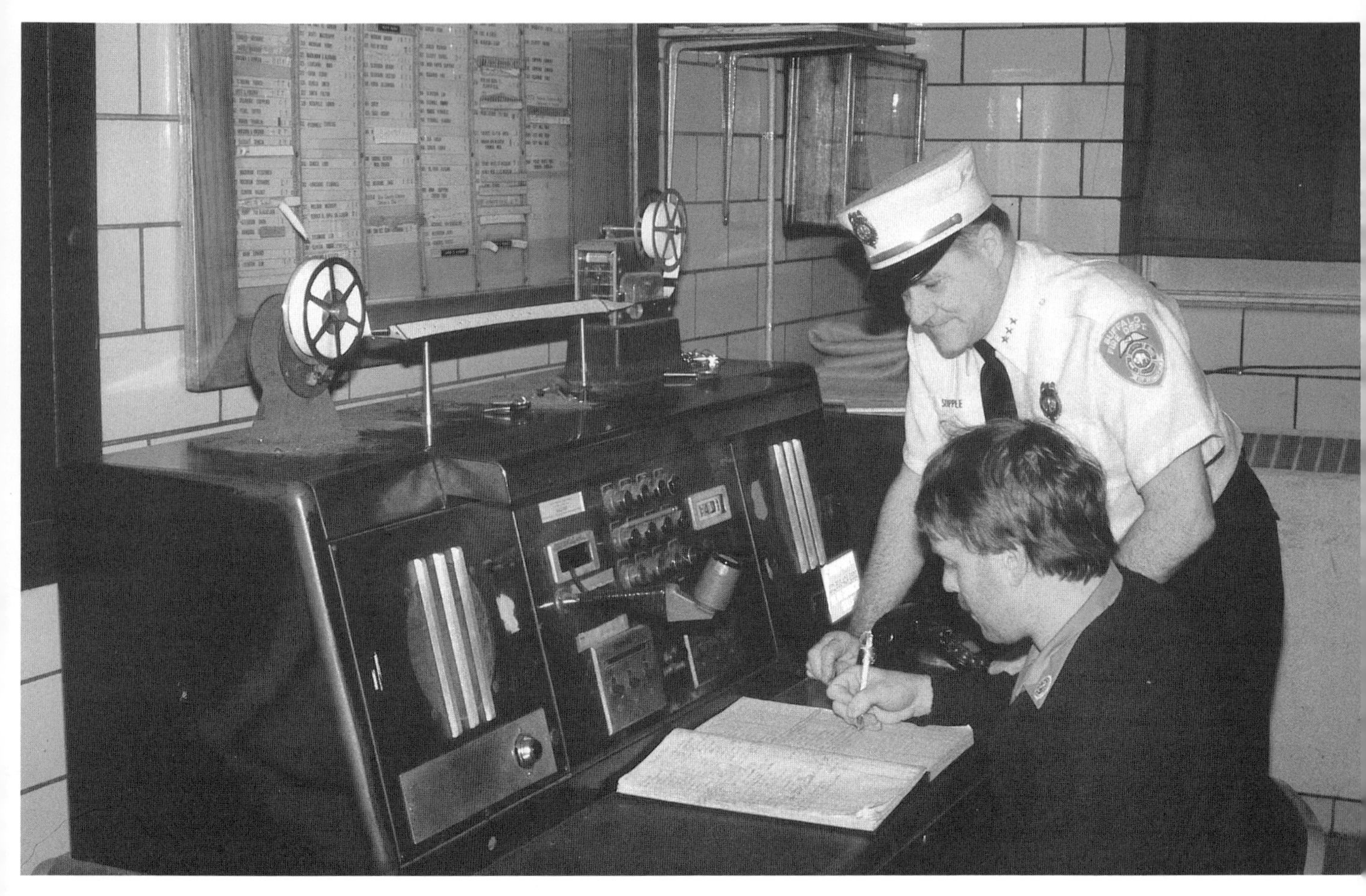

Firefighter William McFeely of Ladder 2, was on housewatch as a box alarm hits. Also counting the holes in the register tape and counting the strokes of the gong is Division Chief Jack Supple. Buffalo Fire Historical Society

Tape continued to weave through the registers as Fire Alarm Box Number 29 was automatically repeated.

The system's built-in redundancy not only insured there was no doubt of the location of the alarm, but alerted firefighters in every city fire station that if the nearest engine and ladder companies could not control the fire, then predesignated additional companies would be called on a second, third, fourth or fifth alarm. The simplicity of the system was evidenced by the assignment card — one for each fire alarm box in the city — that firemen call running cards. A set is in the alarm office and every firehouse.

Clouds of smoke and the glow etching the skyline over the wholesale district told firemen that Box 29 was living up to its mystique. For more than a century, Buffalo firemen associated Box 29 with major fires where death, injury and bizarre occurrences often lurked in the smoke and flames. Firemen called it Hoodoo Firebox 29.

As apparatus converged on Seneca Street they saw the fire was in an old four-story brick building. A man in a long, dark overcoat led Engine 1 and Ladder 2 firemen into the building and directed them to the seat of the fire which they attacked with hoselines.

When Division Chief Jack Supple arrived, the fire was spreading from the first floor up an elevator shaft and mushrooming on upper floors of the Seneca Electric Co. He radioed for a second alarm at 10:58 p.m. Again firehouse bells and alarm registers struck Box 29. This time the number was followed by the two-stroke signal for a second alarm:

Exactly as the running card indicated, the second alarm called out Engines 3, 10 and 2; Hook and Ladders 5 and 9 and Rescue 2.

Supple knew the building was around 100 years old. With its wooden flooring and supports, plus the heavy fireload of merchandise and equipment, the unsprinklered building could collapse as flames weakened its structural integrity. How many times and at how many other Hoodoo Box fires had walls in similar buildings collapsed and killed and injured firemen?

Supple ordered all firemen from the building and called for an exterior attack with nozzles attached to ladders, an elevating tower and a platform. If the walls fell, firemen would be out of harm's way.

Fire and smoke worsened as Supple radioed for a third alarm at 11:02 p.m. Again bells and registers struck the two and the nine signal for Box 29 and this time followed it with the third alarm call: two pairs of three extra strokes. More engines and a hook and ladder were quickly on their way to what Supple saw was going to be a difficult and dangerous fire for the 16 fire companies and 80 firefighters.

As firemen battled the flames, Supple was wary of the Hoodoo jinx. "I have a healthy respect for those old buildings in that area," he said later. And he had plenty of experience and recollections to match that respect. The veteran firefighter not only knew a host of Hoodoo Box stories, but so did his family.

Triangular-shaped holes, punched by alarm registers, point in the direction of the flow of the paper tape. Each triangle represents a part of the number of the box. In this case, Hoodoo Box 29 registered on Buffalo alarm equipment. Jack Supple

Three generations of Buffalo's famous Firefighting Supples knew the bad luck box and what it represented. As leery of any Hoodoo Box alarm was Supple's brother, Harvey, Jr., now a retired battalion chief. Other Hoodoo Box stories had been passed down by their father, Harvey, driver of steam-powered Engine 13, who had answered many an alarm from Box 29 when his pumper was pulled by horses.

After batterings by the large streams boring into the windows knocked down the worst of the fire, Supple felt it was safe to renew the interior attack. Firefighters extended their hoselines and, floor, by floor, bulled the flames backward. The fire retaliated with clouds of smoke and heat as water turned to steam.

It was a long, hard night for the firemen. Freezing weather made footing treacherous as firemen tugged more lines into the building to attack pockets of fire. Ice and slush and the cold wind from off Lake Erie compounded difficulties and perils.

Not until early the next morning was the fire controlled. Once again a Hoodoo Box fire had resulted in severe damage estimated at $175,000. Unlike other Box 29 fires, nobody was killed or injured. Supple was about to log the Seneca Electric fire as just another incident at the bad luck box when a mysterious twist was added to Hoodoo Box lore.

Investigators searching for a cause were told by firemen of the tall man in the long overcoat who had led them to where the fire had started. They assumed he was the night watchman. But they could not find him. If he was the watchman, why did he vanish?

The mystery deepened when Seneca Electric officials arrived and said there was no watchman. The building had been locked when the last employee left work many hours earlier. There were no signs of a break-in. As baffling was the discovery that the dead bolt on the front door was in a locked position. And it was through that door that firemen insisted the stranger had led them. There was no explanation for that.

29 SENECA & ELM; THRUWAY EXIT

ENGINES	HL	Special Equipment				BATT. CHIEF	DIV. CHIEF	DEP. COMM.
1 - 13 - 32	2 - 1	R1				1	DIV	
3 - 10 - 2	5 - 9	R2						
21 - 16 - 18	11							
37 - 33 - 25								
30 - 28 - 19	4							

REVISED 1981

Running card showing the apparatus which answered calls sounded from Buffalo's Hoodoo Box 29. Jack Supple

How did the stranger get into the building? Was he a transient who secreted himself before closing time and later started a small warming fire which spread into a full-blown blaze? If so, how did he manage to leave a locked building? And why would he wait for firemen to arrive and then show them where the fire started? That would almost surely have resulted in his going to jail for burglary and entry, as well as arson. He had every reason, therefore, to vanish into the night before firemen arrived.

The mysterious stranger was never found. The hunt for him ended when further investigation showed the probable cause was an electrical malfunction.

During the year of the Seneca Electric fire, Buffalo firemen answered more than 10,000 alarms, many sounded by passersby who pulled the lever in an alarm box which activated individually coded wheels and other mechanisms that transmitted the alarm by a system of lines to the alarm office. As old as the system was, including some 2000 fire alarm boxes, not one of these boxes had the reputation for treachery that Box 29 had accumulated over the years.

Many have tried and all have failed to explain the bad luck traditionally linked to Hoodoo Box 29 — a jinx unique in the American fire service. The location of the box in an old and heavily-commercialized district is one explanation. Major fires could be expected in these rundown districts where so many buildings were not equipped with sprinkler systems.

Logical as that explanation seems, it fails to take into consideration that there were any other fireboxes in the same area. Sometimes they were pulled for fires. But not one of these boxes compiled a comparable log of death and destruction as did the Hoodoo. The box became so ingrained in Buffalo Fire Department history that two followed by nine bells were tolled in memory of all Buffalo firemen who died in the line of duty. The observance occurred each year during the Firemen's Ball.

The story of the Hoodoo Box began soon after Buffalo installed its first 20 fire alarm boxes in 1866, a year after the Civil War. Buffalo was among the first cities to purchase a system from a company operated by John N. Gamewell in Massachusetts. (Boston, in 1852, was the first.) Buffalo found the system so efficient that more boxes were quickly added, including Number 29 posted at Seneca and Wells Streets.

Box 29, like all boxes of that era, was opened with a key that was issued to policemen and businessmen with property near the box. Use of a key was said to help prevent false alarms. Lack of a key could also result in a delay as passersby rousted store owners or looked for a policeman.

After the door of the box was opened, the alarm was activated by pulling or turning a handle or a lever. This caused the closed electrical circuit system to begin operating: A system of spring-loaded, clock-like gears in the box, including a code wheel with raised triangular-shaped teeth or notches began to turn. Each box's code wheel corresponded to the number of the box. Box 29's code wheel, for

Driver Harvey Supple of Engine 13, a company which answered many Hoodoo Box alarms, backs the rig into the station upon returning from an alarm around 1912. Supple and his sons, Harvey, Jr., and Jack, were to become famous as The Firefighting Supples in the history of the Buffalo Fire Department. The three generations of Supples answered and fought many fires sounded from Hoodoo Box 29. Buffalo Fire Historical Society

Fire alarm box code wheel has teeth or ratchets corresponding to the number of the box. Turning in an alarm causes the electromechanical devices to read these teeth, individual to each box, as the half-dollar sized wheel turns. This code wheel is for Box 351. Fred Conway

example, had two teeth, followed by a small rim, then nine more teeth.

As soon as Box 29, or any other alarm box, was actuated, the code wheel rotated on its gear-driven axle. A small device called a pawl clicked into each valley formed by the ratchet-like teeth. Much like a clock strikes, the electrical circuit was alternately opened and closed as the sprung tension on the pawl caused it to drop into the valley. Each time the pawl dropped, it caused the circuit to open and thereby telegraph an electrical impulse.

This series of two and nine impulses caused an electromechanical bell — also similar to a clock that strikes — to toll the number of the box. In common with communities having similar systems, the bell was centrally-located, often in the tower of a city hall, courthouse or church.

On July 18, 1878, someone turned in an alarm from Box 29 and the bell in a downtown tower tolled the number. As the alarm shattered the early morning tranquility, volunteer firemen hurried to their stations and pulled their apparatus to Seneca and Wells Streets where flames were engulfing the nearby Red Jacket Hotel. As flames soared from the roofs and windows, a brick wall collapsed and killed Fireman John D. Mitchell of Columbia Hose No. 11. His was the first recorded death as a result of a Box 29 alarm.

In 1880, the paid Buffalo Fire Department was formed. The city had, by then, incorporated a fire alarm office in the department's headquarters building. Gamewell, from its Newton Upper Falls, Mass., factory, supplied newer alarm boxes which could be activated by breaking a small glass window on the front door of the cottage-shaped box and turning a key which was an integral part of the box. The alarm was transmitted to the alarm office where it was sent to firehouses by electronic repeater devices. The stations received the alarm on registers mounted on watch desks.

Collage of keys used to open fire alarm boxes to gain access to the alarm tripping mechanism. Steven Scher

Alarm Box keys were issued to police and night watchmen, as well as nearby businessmen. Only they could open the box to turn in an alarm.

Simultaneously, the alarm caused large gongs to clang in rhythm to the strokes of the number of the box.

By quickly referring to a numerical listing, Buffalo firemen knew that, in the case of a Box 29 alarm, five horse-drawn steam-powered fire engines, a fireboat, two hook and ladders, three chiefs and a fuel wagon loaded with sacks of coal answered the first alarm. Even that large assignment of firemen and apparatus was not enough to control what happened around 2:45 a.m., Saturday, February 2, 1889.

Officer Patrick Reardon of Cavanaugh's Night Security Patrol was walking his beat along Seneca Street that bitterly cold night. Swirling snow, driven by a fierce gale-like wind off Lake Erie, made it a miserable night. Reardon had just passed Wells Street when — even above the sound of the howling wind — he heard a muffled explosion. The entire area lit with an orangish glow.

Reardon saw flames spouting from fourth floor windows of the huge Root & Keating Building at Wells and Carroll Streets. The five-story brick factory housed manufacturers of leather boots, shoes, horse harnesses and belts. Running to Seneca and Wells, Reardon broke the glass over the key in Box 29 and turned in the alarm.

Although they quickly arrived, firefighters found flames had extended into the fifth floor and were blowing out of the roof of the factory. Second and third alarms, followed by a general alarm from Box 29, called out nearly all of Buffalo's engines and ladder trucks.

There seemed to be no way to avoid a conflagration. Within 15 minutes, wind-driven flames and radiated heat shattered windows of the gigantic Sherman Jewett Stove Works, one of Buffalo's largest factories. It, too, was soon engulfed.

Confronted by never-seen-before problems caused by the

Buffalo's fire alarm office as it looked in the days of the horse-drawn steam fire engines, ladder trucks and chemical wagons. Many Hoodoo Box alarms were struck from here. Jack Supple

wind-goaded fire, hampered by frozen hydrants and with ice encrusting their apparatus, hoselines, boots, coats and with icicles hanging like braids from their helmets, firefighters could only hope to delay the spreading flames until they got a battleline around the fire. But the fire overwhelmed every stand they made, especially as the wind's fury capriciously changed direction time and time again and sent flames and impenetrable clouds of smoke stabbing north, east, south and west in a five-block area. Worst of all, Wells between Seneca and Ellicott Streets was a solid mountain of fire.

Firefighters narrowly escaped death or injury as roofs, floors and walls of the four, five and six-story brick buildings thundered down all around them. Heat mauled firefighters backward as they took up new stands and were forced to retreat once again. A steam fire engine and a ladder truck were abandoned in their retreat, but firemen managed to save their horses. Apparatus could be replaced. Horses were loved and took a long time to train.

Forty-five minutes after the first Box 29 alarm, the Broezel House and the Arlington, two of the city's best-known hotels, were fully-involved. Guests, awakened by the clanging apparatus, the bellowing flames and the crashing walls, had escaped only minutes earlier.

"Within an hour, the great block of Wells Street was a mass of flaming ruins," reported the *Buffalo Evening News*, "and despite the efforts of firefighters, other buildings in the vicinity quickly caught from the flying fagots of fire."

Among the large buildings lost were the Empire Coffee Mill and the Sibley & Homewood candy factory at the corner of Seneca and Wells. A firefighter later recalled that the candy factory had replaced another building lost in an earlier Box 29 fire. It is likely that the term Hoodoo Box probably originated from that recollection.

The fire, Buffalo's worst since the city was swept by flames in the war with the British and Indians in 1813, raged on hour after hour as more factories, offices, restaurants, stores and other buildings melded into the conflagration despite efforts of firefighters who, said the *News*, "worked like heroes."

Around dawn, the exhausted firefighters managed to surround the flames, but it would be days before all fires were out. Several dozen commercial buildings in the heart of Buffalo's manufacturing and business district were destroyed. Losses were put at $2 million which, even by 20th Century values, makes The Great Seneca Street Fire the worst conflagration in the city's history and one of the worst ever experienced in the United States.

Miraculously, only one fireman was killed: Richard Marion of Engine 10, who was crushed when a hotel wall fell on him. Many firefighters suffered frostbite and other injuries. Severely hurt was Fire Chief Fred Hornung. His arm was nearly severed by a falling shard of a plate glass window. He bore the scar of his Hoodoo Box experience for the remainder of his life.

What is remembered as one of the blackest days in Buffalo Fire Department history started at 5:15 a.m., January 28, 1907, when Box 29 hit on firehouse registers for a blaze in the eight-story brick Columbia Hotel at 101-107 Seneca Street.

Firefighters, jolted awake when the big alarm gongs clanged twice, paused, then clanged nine more times, did not have to look in their running card directory to know where Box 29 was. Arriving, they found that the hotel was puffing smoke and flames. All guests escaped as firefighters quickly laddered the building, attacked with hose streams and called for three more alarms.

Among the best recollections of the fire came from Battalion Chief Lavergne P. Allen, who said it was one of the first of many times in his 31-year career that he had answered a Box 29 alarm. At the time of the Columbia fire he was a fireman assigned to Engine 1, the first pumper to arrive at the fire.

"We were operating a hoseline from the roof of the four-story Burt Candy Company next door to the Columbia when bricks started to fall around us," said Allen. "We dropped our hose and ran for our lives." Allen and his engine company crew escaped, but firefighters on the roof of a two-story building on the opposite side were not so fortunate.

A deadly cascade of bricks and other heavy debris from the collapsing wall drove the roof and both floors into the basement. More than 20 firefighters were buried under the rubble. "We immediately started digging through the debris for the trapped men," said Allen. "I could hear several of my close friends calling for help, but it was several hours before we could reach them. Three were dead." The Columbia fire would stand for many years as the worst loss of firemen's lives in the history of the department.

The disaster escalated superstitious fears of Hoodoo Box 29. A four-alarm fire, September 17, 1913, only added to the jinx legends. Ironically, the Broezel House, rebuilt after The Great Seneca Street Fire, was destroyed, along with the Burnberger Popcorn and Eureka Coffee Companies.

Six years later, the Hoodoo Box heaped more notoriety upon its reputation as Box 29 was sounded for one of the worst and potentially deadly fires in the history of the department.

At 9:15 a.m., July 19, 1919, Chemical Engine 1 answered a call reporting a small fire in the basement of Stoddard Drug Co., 86-88 Seneca Street. The fire was more than these firemen could handle and one of them ran to Box 29 and turned in an alarm for more help. Flames, rapidly spreading throughout the basement, traveled by the elevator shaft and spread into all four floors of the brick building. A second and then a third alarm was called as nearly every piece of Buffalo's fire equipment answered the Box 29 calls.

Fed by explosions of drugs and other chemicals, the fire spread to the Empire Clothing Co., 90 Seneca, a men's furnishings store at 94 Seneca and the nearby Wilson pawnshop. The fog-like smoke put "a haze over the sun for all of downtown Buffalo," reported the *News*.

To put this fire into perspective, Hoodoo Box or not, it must be noted that these were the days before firemen were equipped with breathing apparatus. Hazardous materials, such as the chemicals and drugs in Stoddard's, did not command the resources of modern day firefighters with their special entry suits and protocols of respect and wary fire attacks on lethal toxins created by the stew of assorted and often unknown synergistic effects of dangerous materials.

Toxics hidden in the smoke knocked out 30 firemen as the Stoddard fire burned for three days. Although departmental records show no firemen were killed while fighting the Stoddard fire, we can only speculate on how many lives were shortened by the long term effects of exposure to the tons of burning chemicals in the building. Buried in the *News'* story of the fire was a reference to "the so-called Hoodoo Box."

Nobody knows how many large fires were reported by the sounding of Box 29. Retired Fireman Frank Greene told a *News* reporter who first wrote a feature article on the bad luck box: "I guess I fought 10 big fires that were sounded from Box 29, including several large ones we had at the old Buffalo Paint and Glass Co. on Seneca Street."

Shortly after the *News'* article appeared, an off-duty Buffalo fireman was walking along Seneca Street when he saw black smoke swirling from windows of a four-story building. He knew exactly where the nearest firebox was located because he had answered many fires sounded from Box 29. This two-alarm fire reported from the Hoodoo Box burned a workman in the building and put two firefighters in the hospital after they were overcome by smoke. In the reporting space reserved by the department for noting peculiar aspects of all alarms, the battalion chief scribbled: "Hoodoo Box 29."

As the infamy of Box 29 spread, the National Board of Fire Underwriters looked into the phenomena, along with Robert J. Zahm, who was deputy fire commissioner at the time. The Underwriters could find no other explanation than the fires were purely coincidental to the hazards of the area. As Zahm did, the Underwriters suspected that many Hoodoo Box fires were started by arsonists hired by businessmen to defraud fire insurance companies. This explanation still did not answer the question: Why weren't dozens of other fireboxes in the area culpable, too, for so many deadly fires?

Hoodoo stories persisted over the years, but more of them have been lost than those which are known. The same reporter who wrote the *News'* article went on to become a national magazine writer and book author. In 1955, he told the story of the Hoodoo Box in a national magazine. By that time he was living in Los Angeles and was surprised when a friend sent him a clipping from a September 6, 1955, article in the *Buffalo Courier-Express.* The article said:

"Daniel Bogen, 42, of Engine 8, learned some departmental history from both the written word and actual practice yesterday. Bogen was reading in a national magazine about Buffalo's 'Hoodoo Box No. 29' when an alarm came in from the bad luck box. Bogen put down his magazine and, with the rest of his company, rushed to the blaze."

Shortly after the Seneca Electric fire in 1981, Buffalo finalized the removal of most fire alarm boxes in favor of public and private telephone alarms linked with the 9-1-1 system. As in growing numbers of communities, the little red fire alarm boxes, so commonplace a sight on street corner pedestals, lighted stands and poles, were anachronisms. False alarms, maintenance costs and telephones were largely responsible. With their gradual disappearance from the American street scene, a bit of Americana was lost. New York City remains a holdout. From its more than 16,000 fireboxes, some 37,554 alarms were received in 1989. More than 29,000 of them were false.

The Hoodoo Box, reincarnated again and again as Gamewell improved upon its original system, ended its service life as a 1931 style box with an exterior shell of lightweight Herculite (aluminum alloy) and a white-painted quick-action door. It was activated by pulling down the handle and then a lever.

"I doubt whether too many of the younger firemen on the job know the story of the Hoodoo Box," says Division Chief Supple, who has served the Buffalo Fire Department for more than 39 years. Nor is it far-fetched to predict that one day few firemen anywhere in the United States will have ever seen a firebox in service, much less answered an alarm for one.

Although Box 29 is forever gone from its pedestal and post, its number remains actively buried in the computer software of the Buffalo Fire Department's alarm system. Whenever a 9-1-1 call reports a fire near its old location, the computer selects Box 29. The number will long flash on alarm display screens as a backup for modern-day alarm transmissions by telephone, audio and radio systems.

Officially, Box 29 is known as a Phantom Box. Who could argue with that?

Three Buffalo firefighters were killed while fighting a fire, January 28, 1907, which was sounded from Hoodoo Box 29. Paul Ditzel

THE FIRST FIRE ALARM BOX?

Gamewell literature, fire alarm historians, and even the Smithsonian Institution claim this box to be the first one developed by Dr. William Channing and Moses Farmer. Yet Channing, himself, disputed the wood firebox notion by stating his boxes were cast iron. More than 300 words, attached to the inner door, told how to operate the birdhouse-like box with its quaint wood-shingle roof.
Smithsonian Institution

CHAPTER TWO

"SPIDER WEBS OVER A COW PASTURE"

As with the success of so many other products, the invention and development of the first fire alarm box system came at the right time, the right place and filled the right need.

The man most responsible for the system was not an electronic genius, but an obscure physician who never practiced medicine, but happily and everlastingly was devoted to dabbling in galvanic theories. Dr. William Francis Channing's name is today all but unknown, except to a coterie of fire alarm historians, but it was his ingenuity and persistence — not to mention an amazing talent for huckstering — that resulted in the first fire alarm box and telegraphic system for transmitting alarms.

While many would like to think otherwise and have perpetuated the canard for more than a century, Channing did not do it alone. He had the help of some of the nation's best authorities on clocks, bells, electronics and physics.

It may seem incredible in this age of cybernetics, lasers and electronics that everything from computers and microchips to spacecraft spawns from the minds of highly-skilled and educated engineers and scientists. Channing's invention must, therefore, be put into the context of the times.

The bicycle-building-brothers, Wilbur and Orville Wright, are renowned for their pioneering aircraft flights. Samuel Finley Breese Morse was best-known, in his day, as an artist whose paintings hung in prestigious collections. He enjoyed a wide reputation for his portraits and founded the National Academy of Design. If Morse had done nothing more he would be remembered for his artistic creations.

Then Morse learned of experiments with electric telegraphy. Extending the state of the art, Morse achieved everlasting fame on May 24, 1844, when he demonstrated his telegraph to U.S. Congressmen by tapping the message: "What hath God wrought," over a telegraph line strung along trees and poles between Washington and Baltimore. Broken glass bottle necks served as insulators.

The idea of applying telegraphy to receiving and sending fire alarms occurred to Morse as one of a host of commercial applications. Morse believed the most commercially viable use of telegraph was for financial institutions and railroads. Historians can only speculate that Morse saw greater returns from these and other commercial applications than he probably perceived as a relatively small market for fire alarm systems.

As visionary as Morse was, he apparently did not recognize the critical need for better means of alerting firemen to a problem which began to plague man when he first discovered fire and the consequences for good and evil that were ever after inherent in it. The history of conflagrations is an oft-told story. Nobody knows how much personal and economic suffering could have been avoided had a practical fire alarm system been in place to alert firemen before flames spread beyond control.

THE ORIGINAL PATENT

W. F. CHANNING & M.G. FARMER.

ELECTROMAGNETIC FIRE ALARM TELEGRAPH FOR CITIES.

No. 17,355. **Patented May 19, 1857.**

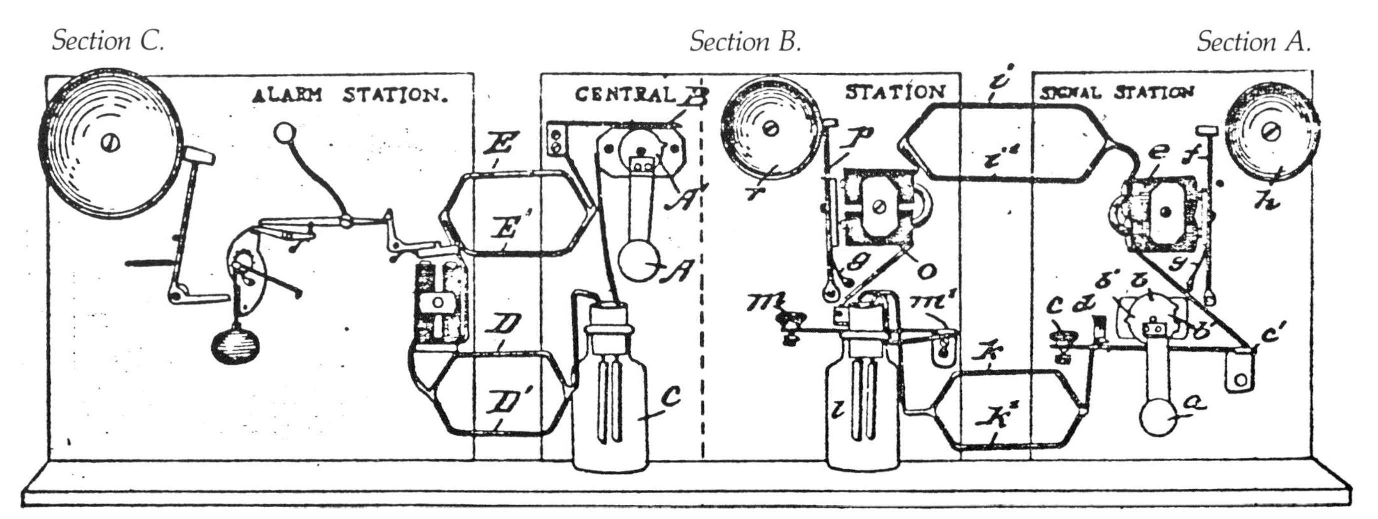

A flagrant example will suffice. A small fire broke out in New Orleans on Good Friday, March 21, 1788. The usual method for sounding the alarm was the same as that in many communities: ringing church bells. But the Capuchin priests, out of reverence for the holy day, refused to allow the bells to be tolled.

By the time any organized firefighting was attempted, nearly 900 buildings were destroyed. Similar to the religious ethos of that time (witness Cotton Mather's preachings after fires in Massachusetts) the priests held that their refusal to ring the bells was justifiable. They took to their pulpits and convinced the citizenry that the flames that laid waste to New Orleans were an act of God to punish people for their sins. There must have been a lot of sinners in New Orleans because nearly the entire city was devastated. It was America's worst fire up to that time.

While New Orleans and fires in Mather's Boston were extreme examples, fire alarm systems in cities — not just American — but throughout the world — were haphazard. Mindful of today's satellites which can, by infrared detection, immediately report the location of forest fires, it is incongruous, indeed, that fire which laid waste to so many cities throughout the history of civilization was so crudely and ineffectively guarded against.

Perhaps an answer lies in the belief — still commonly held — that fire is something that happens to somebody else, never to us. Or the fallacy that fire losses are covered by insurance. That the effectiveness of the action taken during the first few minutes after an outbreak of fire often tells the difference between a trifling fire and a conflagration is idiomatic among firemen.

Methods for sounding fire alarms in colonial times to the start of the phenomenal growth of American cities in the mid-19th Century were a grab bag of sometimes mocking, sometimes incredibly naive approaches: blowing fox-calling horns and conch shells, banging of drums, swinging metal bars in a circular motion while striking the side of metal triangles or banging iron hoops.

Detroit, in 1827, bought a steel triangle formed by bending a 4-inch metal bar. The triangle was eight-feet high and weighed about two tons. Lewis Davenport, generally recognized as Detroit's first paid fireman, was paid $12 to sound alarms. The city's Common Council later voted to pay $5 to the first person who whanged the triangle to sound an alarm. Severe penalties were voted for falses.

The Liberty Bell in Philadelphia is a cherished symbol of America, but it rang more often to call volunteers to fires than it ever did for freedom's cause. Church bells, government building bells and schoolhouse bells were tolled from towers. The first person to spot a fire ran to toll the bell.

Watchmen were common in larger cities which could afford them and needed them because there was so much more of an area to observe. Firemen knew there was a fire when they heard the bell tolling. Glancing up, they saw the watchman pointing in the direction of the fire he observed: Flags by day and lanterns at night. Firemen headed in that direction in the hope that they found a small fire before a large fire found them. In other communities, a red lantern hanging from the tower belfry might signal a fire in the north section of town; a green lantern for a southern area fire, yellow to the east and blue to the west. There were as many variations on this theme as there were fire alarm systems.

Philadelphia's Liberty Bell, prior to its historic tolling, was rung to call volunteer firemen. United States Department Of The Interior, National Park Service

Night patrolmen, discovering a fire, aroused householders by vigorously shaking this wooden rattle. Paul Ditzel

In colonial New York, patrolmen prowled the streets from curfew at 9 p.m. until sunrise. That was the time most fires occurred. Discovering a fire, the patrolmen shook a wooden rattle which aroused householders who threw out leather water buckets they were required to keep on their doorsteps. Volunteers scooped up the buckets as they ran to the fire where they formed bucket brigades.

In Alexandria, Virginia, militiamen fired muskets to sound fire alarms. History fails to show that idea was commonplace along the east coast or even the midwest. Somehow, the idea leapfrogged completely across the United States to Los Angeles where citizens fired pistol shots in the air to call firemen. That was the preferred method for sounding fire alarms in the city even after the paid Los Angeles Fire Department was formed in 1886; more than half-a-century after Morse demonstrated his telegraph and Channing saw his concept of a fire alarm system come to fruition in Boston.

Basic to this hodge-podge of systems was the almost universal failure on any large scale to realize that, before the telegraphed fire alarm box system, the various methods for alerting firemen sent mixed messages. Was the pistol shot signaling a fire? Or just another nightly shootout in Los Angeles? Was the fox horn a call to the hunt or was it for a fire? Did the sounding of a steam whistle atop a glue factory building in Brooklyn mean quitting time or a fire? Were tolling church bells a call to go to worship or a call to go to a fire?

This desperate need for a universally recognized and accepted fire alarm signal exploded as America turned from an agricultural to an urbanized society. In those infant days of what was to become known as urban sprawl, a bell, a hoop, or a triangle could be heard for only limited distances.

This crucial need for a universally recognized fire alarm signal which could be heard over the rapidly-growing square miles of cities was first recognized and implemented in New York City.

On November 28, 1850, New York was divided into eight fire districts. Using the watch tower bells which were installed in 1845, a pre-arranged number of strokes of the bell indicated the district in which there was a fire. A year later, the towers were connected by telegraph; further enhancing the capability of the system to spread alarms faster and, concurrently, the response of firemen. This unifying system no doubt played a role in the organization of the New York Volunteer Fire Department in 1855. New York's alarm system served as a model for others throughout the United States.

Historians credit Charles Robinson of New York as being the first, in 1850, to use the Morse telegraph to transmit fire alarms among these towers and, shortly thereafter, to firehouses. He similarly applied telegraphy to communicate among the city's police stations. Peripheral to the history of fire alarm systems is the growth of police communications via callboxes and related telegraphic devices which closely paralleled that in the fire service.

Also paralleling the start of the first practical fire alarm box system were similarly significant and phenomenal improvements in firefighting apparatus. In 1852, Cincinnati's Latta & Shawk built the first successful coal-fired, steam-powered fire engine. The era of the hand-drawn, hand-pumped fire engines entered their twilight years as even more powerful steam fire engines were built. On the wane, too, were the large city volunteer fire departments. Starting after the Civil War, paid fire departments gradually supplanted the volunteers, except in smaller communities.

A WATCH TOWER OF OLD NEW YORK

Just as 1852 marked the introduction of the first practical steam-powered fire engine, so, too, was the year marked by the installation of the first practical fire alarm telegraphy system in Boston. The system was the first espoused by Channing, a 32-year-old physician who attended Harvard College and received his medical degree from the University of Pennsylvania.

Among historians, Channing is best-known as the son of the famous Unitarian minister and author, William Ellery Channing. Known as "The Apostle of Unitarianism," Channing was ordained minister of the Federal Street Congregational Church in Boston where he served until his death in 1842. Dr. Channing no doubt grew up in the shadow of his illustrious father.

That Channing never practiced medicine indicates the possibility he was nudged into that profession by his famous father. Perhaps of equal significance is the observation that his uncle, Walter Channing, was Harvard's first professor of obstetrics and co-editor of the *Boston and Medical Surgical Journal.* Despite the fact that Walter Channing went on to become dean of Harvard Medical School, his nephew's interests seemed to lie more in the then popular electrical sciences when he received his degree in 1884; another significant year in our narrative because that was when Morse's widely-publicized message, "What hath God wrought," became as familiar to every school student as it did to Channing.

The inner door of Channing and Farmer's crank firebox. The knob protruding from the upper left corner is the shunt which was closed by the pressure of the outer door upon it. The shaft protruding from the slot to the left of the operating handle connects to the signaling lever and permitted its use as a telegraph key for communicating with the Central Office. The knob in the lower right corner controls the tension of the spring on the sounder armature. Smithsonian Institution

If the link between medicine and fire alarm is seemingly impossible to correlate, the explanation can be found in Channing's own words when he presented a paper at the Smithsonian Institute (reprinted in the *Ninth Annual Report of the Board of Regents of the Smithsonian Institute, 1855,* page 153):

"The telegraph, in its common form, communicating intelligence between distant places, performs the function of the sensitive nerves of the human body .... We have ... an 'excitory-motory' system in which the intelligence and motion of the operator at the central station come in to connect sensitive and motor functions, as they would in the case of an individual ..."

Anyone with the least knowledge of the fire service commonly refers to the fire alarm system as "the brain and nerve center of every fire department." Little do any of them realize that the analogy was first made by Channing.

While Channing's classmates became practicing physicians, he pursued his wide-ranging interests, especially if they hinted at adventure and discovery. He was a leader of the first geological expedition in New Hampshire. In 1847 he surveyed the Lake Superior copper region.

Vehemently opposed to slavery, Channing served as an editor of *The Latimore Journal,* a Boston publication named in honor of George Latimer, a runaway slave. Fascinated by electricity in general and magnetism in particular, Channing's medical bent melded with those interests in his 1849 treatise, *"Notes on the Medical Application of Electricity."*

In that specialty, Channing was far in advance of his time, although there seemed to be a bit of snake oil salesman in him. Channing's treatise recommended use of electric shock for the relief of asthma, aneurysms, paralysis and epilepsy. He postulated a gadget for use as an electronic eyewash for patients with poor eyesight.

As dubiously fascinating was Channing's idea for electrically-equipped shoes which would "pass current up one limb and down the other," apparently for orthopedic and neurological problems. To Channing's credit it must be noted that he foresaw a much later time when the electrical and medical sciences would join in various successful protocols.

Not one given to thinking small, Channing appeared before the House of Representatives in 1880 to describe his plan for transporting ocean vessels by rail across the Isthmus of Panama. Many scoffed, a reaction that Channing had long grown accustomed to, but history shows that several years after Channing's death in 1901, ground was broken for what became known as the Panama Canal.

Archives of the Smithsonian Institution say this photo probably is of an experimental model of Channing and Farmer's 1852 fire alarm box. The circuit is open. It closes as the teeth on the code wheel (center) contact the spring, with the wheel actually forming one side of the electrical contact. Historians believe this mechanism must have given extremely poor signals with the small wheel and probably poor continuity as well. Smithsonian Institution

If there are those who would put Channing down as a medical huckster and gentleman adventurer, there is no arguing his phenomenal ability to garner publicity. Unlike many doctors, inventors and persons of science who tend to be introspective and shy of public attention, Channing was an expert at hyping his ideas as a public speaker and writer. No publicity agent was more skilled than Channing at putting his name and ideas before the public.

Channing wisely decided that if his concept for a fire alarm system was to become reality, he must have public support for the costs of building it. Politicians, ever sensitive to the electorate, could therefore be more easily convinced to vote the necessary financial outlay.

Bylining an article in the *Boston Daily Advertiser*, June 3, 1845, Channing described his proposed system of fire alarm boxes, overhead telegraph wires strung between poles and leading to police and a central fire alarm office where these alarms would be received and transmitted to firehouses.

Although Boston was his home, Channing could not have chosen a better city in which to make his pitch. Boston, more than any other American city, had suffered an epidemic of major fires from colonial times in the city mostly built of wood. The citizenry — and their elected officials, so Channing believed — were eager to hear of anything that would diminish the threat of conflagrations; an ongoing terror that can only be likened to that of earthquake and fire prone Californians starting in 1906 with the calamity in San Francisco.

Channing's expectations and contemporary accounts to the contrary, Boston officials were not as enlightened and eager to respond with backing. Not to be deterred, Channing continued his sales pitch wherever anyone would listen and anywhere anyone would publish his ideas.

While Boston waffled, Channing realized he was long on theory and short on execution. Convinced that if Boston did not proceed, some other city would, Channing determined to be ready to actually build the system. He partnered with Moses Gerrish Farmer, 32, a native of Boscawen, N.H., who had pioneered the development and maintenance of telegraph systems in Massachusetts. Farmer, who lived in Salem, Mass., was, according to a contemporary account, "the most expert electrical mechanic of the day."

Practical as he was, Farmer was a dreamer similar to Channing. Among Farmer's inventions was an electric train that carried children and a process for electroplating aluminum. He invented a dynamo which he used in experiments to develop an electric light bulb. He may have become famous for that, except that Thomas Alva Edison beat him to it by inventing the first practical light bulb in 1879. What most intrigued Channing, however, was Farmer's 1848 invention of a bell striking device. It would find its way into the Channing & Farmer fire alarm system.

Boston finally yielded the political impasse in 1848 when Mayor Josiah Quincy, Jr., urged city councilmen to at least investigate the potential for Channing & Farmer's fire alarm system. The council voted funds for construction of two of Farmer's bell-striking machines that would strike city bells from distant points.

In his *History of Boston Fire Department*, John Dale & Co., 1889, Arthur Wellington Brayley wrote: "These were constructed under the supervision of Farmer and one of them was placed in the belfry of the old City Hall and was connected with a line of telegraph wires extending to New York City, where the operator, following his instructions, opened and closed the circuit by means of his 'key,' resulting in a series of blows on the bells in this city, which according to the papers of that date, 'caused a false alarm of fire.' "

We can but speculate that Boston was hedging its bet by sending false fire alarms from New York City to Boston. If the fire alarm system proved to be a flop, they would at least have better telegraphic communication with New York.

Channing and Farmer apparently went through much experimentation before developing a satisfactorily operating fire alarm box. This photo, believed to be of an experimental mechanism, shows a larger code wheel than in earlier models and is turned by a gear instead of a crank. Note the ratchet on the small gear (above and to the left of the crank pivot) that prevented a person from turning the crank to the left. In this model, too, signaling is accomplished by an arm that rides on the code wheel instead of using the wheel itself. This mechanism apparently was for use on a closed circuit, although the contacts are missing. Smithsonian Institution

In 1851, three years later, the city appropriated $10,000 for construction of Channing & Farmer's fire alarm system, "a decision much to their honor, as it was an experiment, being without precedent in the world," said Brayley.

(Some would quarrel with that statement. Berlin, Germany, in that same year installed 46 signal stations which were connected with each other and a central station for the use of police and firemen. The German system, built by Siemens & Haskie, was referred to as a dial-telegraph. U.S. Patent Office investigators later said this system was unlike Channing's or one in New York, for that matter, which was a Morse telegraph code system. There appears to be no doubt, therefore, that Channing rightfully deserves credit for his

The first Central Fire Alarm Station was located in Boston. Note the piano-like transmitting keyboard in the center of the illustration. The three instruments directly above it registered the number of times boxes on three circuits were pulled. As the pointer on the gauge lowered, it indicated it was time to rewind the boxes. According to contemporary descriptions, the telegraph lines leading from the station to the alarm boxes resembled "Spider webs over a cow pasture." Robert Fitz

concept of the first practical and successful fire alarm box system. To Farmer must go the credit for supervising its construction.)

Additional perspective comes from a lengthy article in *The Commonwealth,* published in Boston, December 30, 1851:

"The observing must have noticed of late a great increase in telegraph wires stretching over the housetops of the city in all directions, like spider webs over a cow pasture in a June morning. In the vicinity of the City Hall this is especially noticeable. If one takes his position in front of that building and looks upward, the air seems to be full of wires which converge to a centre (sic) over a building ... where the great municipal spider seems to have its nest.

"Such indeed is the fact, for in the attic of that building, belonging to the city, the soul of the whole arrangement is to reside. This arrangement is one of the wonders of the age, involving in it one of the deepest mysteries of the universe; and the more it is studied the more mysterious it becomes ..."

Equally mysterious are the various descriptions of the first fire alarm boxes, related telegraphy and how the system worked. Because the system has been retold so many times we must turn to Channing's own description, *"On The Municipal Electric Telegraph, Especially In Its Application to FIRE ALARMS,"* published by Yale College in 1852 and extracted from *The American Journal of Science and Arts* (Volume XIII, Second Series).

Channing's description of the fire alarm boxes:

"The signal instruments are contained in a strong cast iron case and connection is made between this and the conductors on the top of the building by a wrought iron tube enclosing insulated wires ... Boston ... wisely decided to place these boxes on the outside of buildings, in places well selected, generally opposite a lamp.

"These stations are distributed throughout the city at distances not greater than 100 rods from each other, so that no house shall be distant more than 50 rods from one of them," says Gamewell, who says there were 41 fireboxes. Other sources agree that the boxes were painted black and that the system included about 49 miles of wire.

Gamewell continues: "The signal box and door consist each of a heavy casting .... The outside of the door has upon it the words, Signal Station, with the number of the station and a panel containing a notice of the place where the key is to be found .... The signal crank with a heavily-weighted handle is within the box. It was devised by Farmer and myself to obviate the irregularity which might arise from the manipulation of the signal key by ignorant or incompetent persons ...

"The axis of the crank carries a circuit wheel provided with a number of teeth or cams, each of which, revolving, completes the circuit momentarily by a sliding contact with (another) key. These cams are divided into two groups ... one on each side of the circuit wheel, the principal of which groups numbers from one to seven cams, according to the number of the district in which the signal box is placed.

"This communicates the district number to the central office. The other group consists of from two to four cams, placed closely together, and so formed as to complete the circuit for longer or shorter periods and produce a record at

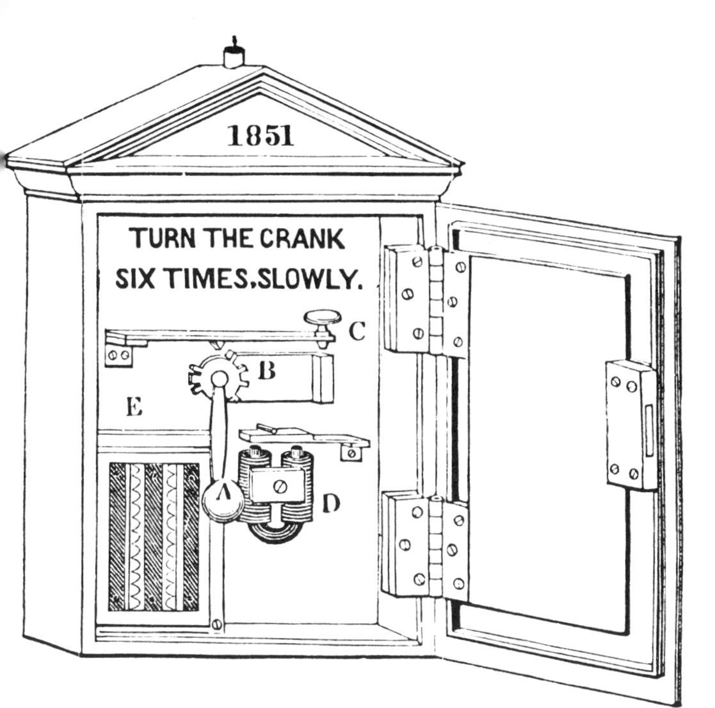

Dr. William Channing provided this illustration of what he called his first fire alarm box. It was made of cast iron and apparently was 16-inches tall, 11-inches wide and 4-inches deep. Channing alphabetically described the interior mechanism of the box: "The Signal Crank, with a heavily-weighted handle is seen within the box (A) ... The axis of the crank carries a circuit wheel (B) provided with a number of teeth or cams, each of which, in revolving, completes the circuit momentarily by a sliding contact with the key (C) ... The Signal Key (C) can be used in the ordinary manner to communicate to the Central Office any system of signals which may have been agreed upon for police or other purposes. Communications may be received, in return, from the Central Office, by means of the little electromagnet and armature (D) ... The Discharger of atmospheric electricity is represented at (E)." Paul Ditzel

the central office of dots and lines, indicating the number of the station.

"The box contains instructions to turn the crank six times," says Channing. This signal was telegraphed to the central office where it is registered and indicated on electro-mechanical devices. "The object of the repetition," says Channing, "is to draw attention and ensure its correct reception."

Above and to the right of the alarm crank was a signal key "used in the ordinary manner to communicate with the central office any system of signs which may have been agreed upon for police or other purposes," said Channing. "Communications may be received, in return, from the central office by means of the little electromagnet and armature" in the box. "The click of the armature constitutes here the audible signal."

The lower left corner of the box contained electrical grounding. "It consists of three strips of brass, resting on varnished wood and covered with a glass plate, with strips of India rubber cloth interposed," said Channing.

"Each signal station is in charge of a person or family in the immediate neighborhood, whose duty it is to open the box in case of an alarm and turn the crank. This act is so simple that it might be performed by a child. Certain members of the Fire Watch and Police Departments are also provided with keys to the signal boxes."

Channing was perhaps naive in his assumption of the simplicity of the box's operation. This and the subsequent history of fire alarm technology would clearly demonstrate that unforeseen circumstances could lead to a bungled alarm. The crank in Channing's early boxes apparently could be turned in either direction. Only a clockwise turn sent an alarm. In the understandable excitement following an outbreak of fire, the person turning in the alarm could turn it counter-clockwise or turn it so fast as to result in a garbled alarm.

Various contemporary accounts indicate these problems were at least mitigated by retrofitting the boxes with ratchet devices which prevented counter-clockwise turns. Still another contraption apparently prevented the lever from being turned too fast. More than one authority says the early instructions of six turns of the crank were later changed to 26 cranks so the person receiving the alarm in the central office could adjust his equipment to synchronize with an alarm being sent too fast.

Gamewell's description continues: "The centre of the system in Boston is established in the City Building adjoining the City Hall. From its roof which is isolated, the wires, elevated on a bracket, radiate in all directions.

"The instruments at the central office are in part receiving and in part transmitting, besides the batteries for the whole system and the testing and registering instruments employed in the regulation of the circuits.

"The receiving apparatus consists of an alarm office for each of the three signal circuits and a single electromagnetic register of the Morse construction with which they communicate in common.

"The register used is the common electromagnetic regis-

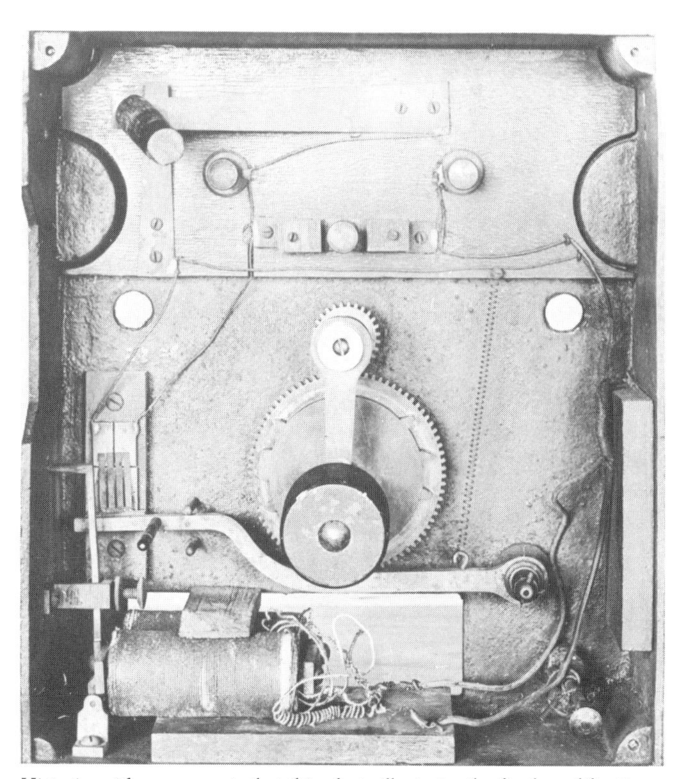

Historic evidence suggests that this photo illustrates the final modification of the inner portion of Channing and Farmer's alarm box. There is no apparent ratchet on this box; probably because it did not require it as the box would signal the correct number when turned in either direction. The photo leaves no doubt that this box was made of cast iron. Smithsonian Institution

ter, arranged so as to start and stop itself. It is made to run faster than usual, so as to record legibly the signals made by the signal crank, even when turned rapidly. The register may be operated by the same local circuit as the three office alarms .... The alarms connected with the different circuits are provided with bells of different tone, so that it is immediately perceived, by the sound, from which circuit any signal proceeds."

Turning to the alarm transmitting device, which was connected with the alarm circuit, Channing said it consisted "of a single signal or rather alarm key and of the district keyboard." This miniature organ-like device was fronted with a row of keys, or buttons, which, when pushed down, activated the transmission. The transmitter was equipped with push-button arrangement for transmitting an "All Out" signal which told firemen still on their way to the fire that they could return to their stations.

Church, school and firehouse bells were part of the system. Channing's description of these bells is not only fascinating but leaves little doubt that the clanging of at least some of the bells could, to coin a cliche, wake the dead:

"The bells to which the striking machines are applied in Boston vary in weight from 3700 to 300 pounds. The machines are of uniform size, but they are carried by weights, varying from 2000 to 800 pounds on a single chain. It was supposed at the outset that a blow equal in force to that of the common tolling hammers would be sufficient for all the purposes of an alarm, especially as, in the telegraphic system, an alarm is not propagated by sound from bell to bell, as in the ordinary method.

"A greater amount of sound was, however, considered desirable by members of the fire department and a great addition to the force of the hammer was found necessary to produce adequate vibration in the largest bells. Thus the hammer, judged suitable for the bell of the Brattle Street Church, weighs 40 pounds, has a handle 3½-feet-long, swings through an arc of 4½-feet and is moved at each blow by a force equal to a weight of 1440 pounds falling one inch."

Channing goes on to explain that the average number of blows in striking a district signal "with intervals of five seconds after each signal, is about 20 per minute. From 50 to 100 blows would be sufficient, ordinarily, for a single alarm." Channing's lengthy paper mentions a great deal of peripheral equipment, including gauges which kept count of the number of strokes and thereby indicated when it was time to rewind the clock mechanism which struck the bells.

By far the most revealing description of the system is Channing's step-by-step account of a typical alarm:

"A fire having broken out in the neighborhood of the fifth signal station in the fourth district, the person in charge of this station, or at night, a watchman, opens the signal box and turns the crank six times.

"The alarm at the central office is struck every time that the circuit is closed and the register records, at the same moment, the district signal of four consecutive marks, six times repeated, alternating with the telegraphic signal, a dot, a line and a dot, indicating the number of the fifth station.

"The agent at the central office, if aroused at night by the alarm, refers to the register where he finds a distinct and permanent record. He turns immediately to the district keyboard and depresses the key of the fourth district.

This keyboard was the original transmitter when the first fire alarm box system went in service in Boston on April 28, 1852. The dispatcher had the option of which key or keys to depress to transmit an alarm. The alarm continued to ring in boxes and belfrys as long as the key was depressed. A separate key was reserved for transmitting a signal that told firemen the blaze was out and that no further assistance was required. Smithsonian Institution

"The battery is at once thrown on to the alarm circuits and the signal of the fourth district, *one, two, three, four,* is struck upon the (city's) 19 alarm bells at nearly the same instant of time and continues to be repeated at short intervals as long as the key of the district is held down.

"The agent, meanwhile, observes the motion of the numerical cylinders in one of the alarm bell registers and raises his finger from the key when a sufficient number of blows have been struck. He then turns to the Journal of the office and enters the time and the number of the district and station from which the alarm (was received).

"In the meantime, the engines are running from all quarters towards the district and some officer of the fire department, wishing to know the number of the station nearest the fire, opens one of the signal boxes in passing and makes the most simple signal, say, *one, one, one* or 'writing dots' by tapping on the signal key.

"This is received by the central agent, who proceeds at once, by means of the key provided for that purpose, to count off the number of the station originating the alarm, on the electromagnets in all the signal boxes of the circuit through which the inquiry is made.

"The engines are thus directed to the exact part of the district from which the alarm proceeded and they should be further guided by a map of the city, prepared for that purpose, with the numbers of the stations and districts marked upon it. If the number of the station as well as of the district should be struck primarily on the alarm bells, any inquiry would, of course, be rendered unnecessary and a direction would be at once furnished to the place of the fire within the distance of 50 rods.

"At length the fire is suppressed, perhaps in a short time.

Boston's first transmitting keyboard had a hinged top. When raised it permitted adjustments and maintenance of the electrically-operated keys and other devices. Smithsonian Institution

A very important function of the system is now to be developed. The engineer on the ground, who has chief control, sends to the nearest signal box and communicates the signal, *one, one-two, —, one, one-two,* which signifies 'All Out.'

"This is received by the agent at the central office, who immediately depresses the key of the district keyboard" used for the sole purpose of striking the notification that the fire is out. "This signal is forthwith struck and repeated a few times on all the bells. The engines in various and perhaps distant parts of the city turn back."

The question obviously arises: What would firemen have done if the entire Boston Fire Department could not control the fire? There was, at that time, no provision for second, third or fourth alarms. As history has so many times shown, the telegraph was used to call for help from neighboring cities. Sometimes help came from great distances.

At the height of The Great Chicago Fire in 1871, for example, Mayor Roswell B. Mason ordered this message telegraphed: "Chicago is in flames ... Send help." Within hours, more than 25 engines from eight states neighboring Illinois were speeding to Chicago via railroad flat cars. Before they got there, more help came from communities around Chicago.

As can be imagined, there was great public interest among Bostonians as their skyline was etched with wiring and fire alarm boxes were attached in prominent areas. If the colonial ambience of the old city was being blighted by this maze of wiring, there were doomsayers who doubted the system would work. But nearly everyone agreed necessity outweighed the ugliness of what *The Commonwealth* appropriately called spider webs. On April 28, 1852, the system was officially turned over to Boston and it was declared in service.

Everyone awaited the first fire alarm to be sounded from one of the signal boxes. It was not long in coming. At 8:25 p.m. the day after the system went in service, J.H. Goodale saw a fire in a wood framed barber shop and ran to the Cooper Street Church where District 1, Station 7, was mounted. In his excitement, Goodale turned the lever faster than the system could register the alarm. Failing to hear tolling alarm bells, Goodale hurried to the central fire alarm office to report the fire which was small.

Meticulous as Channing's plans were drawn, there was a large gap between what was said on paper and what happened on the streets of Boston. Refinements were quickly implemented. Among them was the elimination of the need to go to a district box and tap out a request for the number of the box whose crank had been turned. Both district and box number were simultaneously transmitted.

Only time and experience would identify glitches which would need to be addressed and solved before the fire alarm box system could achieve the sophistication it ultimately did.

John Nelson Gamewell, 1822-1896

A South Carolina postmaster and telegraph company agent, John Nelson Gamewell founded the company whose name was to become synonymous with fire alarm boxes and fire alarm telegraph systems throughout the world. Although there are at least 36 other known manufacturers of fire alarm telegraph equipment (see Appendix D), Gamewell held a virtual monopoly, with a 95% market share.

CHAPTER THREE

THE SMITHSONIAN STRANGER

Despite the many bugs to be worked out in Boston's fire alarm system, it attracted widespread curiosity, especially among entrepreneurs seeking entry into or already dabbling in the infant marketplace for telegraphy.

Realizing that what could be made to work in Boston offered financial incentives elsewhere, the system's developers, William F. Channing and Moses G. Farmer sought to protect their ideas. On May 13, 1854, they applied for a patent. Curiously, an examination of that Patent No. 17,355, shows that the application, as granted, would be wholly held by Channing.

Although Farmer remained in the employ of Boston while supervising operations and improvements in the system, we can only wonder why he turned over the rights to Channing. Perhaps we will never know. As unexplainable is the observation that although two years had passed since Boston's system went in service, no other city had copied the system prior to the patent application. A probable explanation was the well-known fact that Boston's system was not altogether failsafe.

Whether from ego or as a possible marketing strategy (possibly both) Channing arranged to deliver a lecture, *"The American Fire-Alarm Telegraph"* at the Smithsonian Institution in March, 1855. The lecture was well-attended by many businessmen, especially those with a strong interest in telegraphy. The fact that Channing's presentation would be reprinted in the *Ninth Annual Report of the Board of the Smithsonian Institution,* 1855, assured him even greater attention.

Among those in the audience was a 33-year-old portly and balding South Carolinian, John Nelson Gamewell. He was little known outside of Camden, South Carolina, where, in 1843, he replaced his ailing father-in-law as the community's postmaster and agent-in-charge of the local office of the Washington and New Orleans Telegraph Company. A bookseller by trade, the new job was much to Gamewell's liking because he was an amateur telegrapher. Now he could receive pay for his hobby.

Little did anyone in that Smithsonian auditorium — much less this amateur telegrapher who had few friends outside Camden — dream that the Gamewell name soon would become synonymous, even generic, to police and fire alarm telegraphy. Nor that Gamewell would ultimately corner 95 percent of the police and fire alarm telegraphy business in the United States and become the alarm system of choice throughout the civilized world.

There is no indication that Gamewell, a man of modest means, journeyed to Washington with any thought of entering the fire alarm business. In the first place, he did not have the capital to do it. Gamewell's chief claim to fame at that time was that he was the son of the late Frank Asbury Gamewell, a pioneering Methodist minister who was well known throughout southern states.

Gamewell, one of nine children, was four when his father died. Without a father figure, he appears to have been introspective and what would today be described as a loner during his early years. Perhaps this largely explains his later propensity for promoting the name Gamewell, although many technological improvements in fire alarm systems were developed by others.

A superb public speaker, Channing's presentation was all and more than Gamewell expected. With each new thought, Gamewell's enthusiasm heightened. Channing told how Boston's annual fire losses had, because of the instantaneous, universal and definite operation of the system, resulted in faster response times for firemen and a resulting decrease in large fires.

Boston's annual fire losses had dipped to "less than one dollar for every inhabitant; a loss which, for its small amount and in so compact and wealthy a city, cannot be paralleled in America," said Channing. With very little extrapolation, Gamewell realized the significance that this could have upon lower fire insurance rates and what a strong sales message this carried when presented to city officials everywhere.

Gamewell was all the more won over as Channing concluded his lecture:

"The mechanism of the fire telegraph is arranged and disposed for the purpose of preserving wealth, the fruit of human industry and of nature's bounty, from destruction. It therefore accomplished an end of human use. But more than this, it is a higher system of municipal organization than any which has heretofore been proposed or adopted. In it the New World has taken a step in the forms of civilization in advance of the Old."

Returning to Camden, Gamewell sought out his friend, James Dunlap, a well-to-do jeweler and merchant who agreed to financially back Gamewell's enthusiam for purchasing rights to the system for installations in the southern region of the United States and contiguous areas, including St. Louis, Mo., where he had family connections.

Gamewell's meetings with Channing went well. They had much in common: They were about the same age, stemmed from strongly religious backgrounds, were opposed to slavery and, most of all, shared omnibus interests in things electrical. Gamewell was, himself, something of an inventor and described his lightning arrestor which enabled messages to be telegraphed during thunderstorms.

Gamewell, with Dunlap's financial backing, purchased rights for marketing Channing's system in the south and southwest in 1855 and four years later purchased all rights for around $30,000. Why Channing did not remain to further share in the bonanza which was to become the fire alarm equipment industry and was to number around 36 manufacturers, is a matter of speculation.

James M. Gardiner, 1819 - 1915. A clockmaker by trade, he joined his brother-in-law, John N. Gamewell, in the fire alarm business in 1856. He remained with the Gamewell firm until his death at age 95.

When John Gamewell was destitute after the Civil War, Gardiner loaned him $5. Gardiner invented a fire alarm box, which Gamewell named after him. His many inventions and patents for fire alarm telegraphy contributed significantly to the Gamewell success story.

Some theorize that the volunteer firemen saw these systems as a wedge which would ultimately lead to full-time, paid fire departments. They might be expected to fight fire alarm systems. If this theory is to be believed, then Channing did not foresee the dividends that Gamewell did. While there may have been some disgruntlement among the volunteers, the history of their service tends to demonstrate otherwise. Volunteers were known to constantly be alert for better means of receiving and getting to fires. Pride and sometimes bonuses for being first to arrive are part of the lore of their rich history.

It seems more likely that Channing, as was his nature, did not long dwell on any particular electrical device. We can only surmise, therefore, that, as with his myriad other ideas, including the sense of adventure and exploration, his brain was such a fountain of ideas that one heaped onto another. Channing was a dreamer, not a shrewd entrepreneur. Gamewell saw the financial opportunities and foresaw a national and an international market which could be tapped by his inherent business opportunism.

Gamewell was not an overnight success and came perilously close to ruination. Up to 1861, Gamewell managed to sell only a handful of systems in Philadelphia, St. Louis, Baltimore, New Orleans and Charleston, S. C. The ruination factor entered with the outbreak of the Civil War. Guns, ammunition and protection of cities, especially those in the south, became paramount as General William Tecumseh Sherman's troops steamrollered during the infamous march into Dixie.

With Camden in the direct path of Sherman's onslaught, Gamewell, loyal South Carolinian that he was, served the Confederacy in a number of positions, including First Assistant to the War Secretary of the Independent and Sovereign State of South Carolina. Too, he operated the Nitre (or Saltpeter) Works near Columbia, which manufactured gunpowder ingredients. Obviously, this was a target of Sherman's raids. Gamewell fled to Camden where he lived in a swampland until Sherman's forces wrought their devastation and continued their march to Georgia.

After the war, Gamewell found himself destitute. The U. S. Government had confiscated his fire alarm patents and auctioned them in May 1863 from the steps of the Camden City Hall. Virtually penniless, Gamewell was helped by his brother-in-law, James Gardiner of St. Louis. Although Gamewell had nine children, not one of them was to join him in the fire alarm business. But Gardiner's name must be marked. He not only gave Gamewell $5 when he did not have a dime to his name, but was to play a vital role in the development of Gamewell's fire alarm system.

In a scenario that was to become fairly typical during the course of Gamewell history, Gardiner supervised the construction, operation and maintenance of St. Louis' first system. Technological improvements in the original Gamewell system were often devised by city-employed supervisors in communities with alarm systems. These, coupled with Moses G. Farmer's many contributions, were the true power behind the name Gamewell, who was a consummate businessman.

Gamewell sold the systems and their improvements to communities while those responsible for these improvements largely remained anonymous. An indication of Gamewell's sales acumen is shown in St. Louis' initial system including 45 fire alarm boxes costing $23,000, more than double that which Channing received from Boston.

Like so many other southerners who lost all or nearly all as a result of the war and its harsh aftermath, Gamewell's future appeared to be grim. Dunlap, his financial backer, was disgusted and withdrew any further thoughts of starting afresh. He could hardly be faulted. Charleston had been burned on December 11, 1861, by Sherman's troops and the rest of the state, especially urbanized areas, were severely distressed economically, politically and socially.

Recognizing that the Old South was no more, Gamewell, in 1866, moved his wife and family to Hackensack, New Jersey. For anyone facing the adversity that Gamewell did, there was scant optimism over life's cyclical nature. If the pendulum swings poorly, it inevitably returns the other way and better times result. In Gamewell's case the pendulum swung his way when John F. Kennard of Boston went to Washington, D. C., with wherewithal to pay $20,000 for Gamewell's confiscated patents.

For reasons that perhaps will never be known, Kennard purchased the patents for the pittance of $80. Returning most of them to Gamewell, the firm of Gamewell, Kennard & Co. was formed in 1867 in New York. Manufacturing facilities were located in Upper Newton Falls, Massachusetts.

With the end of the war and the phenomenal economic growth in northern cities, interest in better fire protection ranged from fireboats to fire alarm systems. More companies entered the fire service marketplace, notably that of Charles T. and J. N. Chester. In the year that Gamewell and Kennard formed their partnership, the Chesters marketed a better fire alarm box which included a faster-acting lever, rather than the standard crank handle.

On August 1, 1869, the Chesters picked the ripest plum of all when they sold their system to the Fire Department of New York, which had become a paid department in 1865.

Gamewell Umbrella-style tapper bell for sounding alarms in the home or office of commissioners or chiefs of the Fire Department of New York around 1915. Steven Scher

This multi-circuit ink and paper alarm register was in the Mercer Street Central Office in New York's Manhattan around 1870. The device recorded incoming box alarm signals on a roll of paper. Steven Scher

Switchboard and box alarm receiving registers in Boston's Bristol Street Office. Note the sound amplifier at the extreme left of the photo. William Noonan

Later history would show, as New York went, so did large numbers of other cities. The cachet of the Fire Department of New York was a powerful inducement for many fire departments to emulate New York in apparatus, fireboats and alarm systems.

Starting at this time, the history of fire alarm technology weaves a tangled skein. Patents by Chester and others to the contrary, fire alarm system producers commonly infringed. The precise reasons are lost. Perhaps the era was not as litigious as we would later see. The business had an incestuous element to it as inventors of better devices stood by helplessly while their ideas were freely incorporated into the equipment of better-known companies. Realizing that if they could not beat Gamewell, they would join him, many of these smaller companies — including the Chesters — were acquired by Gamewell.

Still another possible explanation is the recognition that every community, however large or small, found a fire alarm box system essential. Maybe the numerous producers figured there was more than enough business for all. A 1904 U. S. Census Bureau compilation is illuminating:

During the decade from 1852-1862, only four systems were fully-installed. This leaped to 40 systems from 1862-1872 and jumped to 62 systems from 1872-1882. As industrialization and other growth, with its concurrent fire problems mounted in cities, installations skyrocketed to 299 from 1882-1892 and another 359 from 1892 to 1902. After 1904 the Census Bureau reported some 764 fire alarm systems in operation, including 37,739 fire alarm boxes.

Despite the rapid growth of telephone services, the same Census Bureau report strongly suggested that alarm boxes were the alarm system of choice. Perhaps because so many families still shared party lines with other subscribers (and might not surrender the line in event of an outbreak of fire) the Bureau noted:

"It has been the practice of the Wisconsin Telephone Company of Milwaukee to suggest in its telephone directory that patrons send in fire alarms by telephone. The Chief of Police has lately requested the Manager of the Company to omit this suggestion from the book hereafter, for the reason that it frequently takes too long a time to notify the fire headquarters by telephone. This delay, he states, gives the fire a chance to gain headway before the department is able to respond to the call."

Gamewell quickly cornered the market in those days prior to anti-trust, anti-monopolistic legislation. The astute

THE "HOUSE-WATCHMAN" AND THE GONGS. *Steven Scher*

Gamewell was to coin slogans which became part of catalogs, advertisements and other publicity: "A Box a Block"..."A Box on Every Panic Spot." In capturing around 95 percent of the total fire alarm business, the name Gamewell became as generically associated with fire alarm equipment as the soft drink, Coke, was to Coca-Cola. And just as jealously guarded.

As fast as testimonials, mostly requested, flooded into Gamewell's office, they were quickly reprinted and circularized throughout the fire service in the United States and other countries. A few samples:

Albany, New York Fire Chief James M. McQuade:

"We would much prefer to have four steamers with the telegraph than eight steamers without it; and the same will hold good in any city."

Chief Engineer E. G. Megrue of the Cincinnati Fire Department:

"Our city has been using Gamewell & Co.'s American Fire Alarm Telegraph for the past eight years. It is impossible to fix an estimate of value, but it more than pays for itself every year."

Chief Engineer A. P. Leshure of the Springfield, Massachusetts Fire Department:

"The fire department are (sic) on the way to the fire within one minute after the first stroke of the gongs."

A keyless type of fire alarm box introduced in New York City during the 1880s. Steven Scher

Fire Marshal Mathias Benner of Chicago:

"I believe the sum per station saved annually to the city (in fire losses) to be at least $2,500."

Chief H. C. Sexton of the St. Louis, Missouri Fire Department:

"I consider it (the fire alarm system) as the right arm of any fire department."

In Hartford, Connecticut, insurance officials jointly wrote Gamewell:

"This system of fire alarm has been in operation in our city for the last three months, during which time there is no doubt but what it has already saved its cost, in one or two instances of fire."

As quickly as Gamewell widely spread their latest testimonials, other fire alarm companies fired back with barrages of their own; obviously intended to denigrate Gamewell's equipment.

Several of them tend to balance Gamewell's hoopla, if we are to believe these comments repeated by the United States Fire and Police Telegraph Company, Boston, which published them in an 1897 booklet shaped like a miniature fire alarm box:

Chief Henry A. Hall, Methuen, Massachusetts, Fire Department: "We have had more trouble with the Gamewell in one storm than with all the U.S. (boxes) in the two years we have them."

Chief Robert Holman, Portland, Oregon, Fire Department: "We have used the Gamewell system since 1874, but have adopted yours as we think it is a great improvement over all others."

Fred M. Vandevoort, Assistant Superintendent, Fire Alarm, Omaha, Nebraska: "The U.S. Successive Box is the *only* fire alarm box."

Behind these little red fire alarm boxes which seemed to sprout on poles and pedestals throughout cities, were the many unknown technicians who made these systems possible. Perhaps the best way to trace these developments is by means of a Fire Alarm Almanac (Chapter Four) and a Cavalcade of Fire Alarm Boxes (Chapter Five). In addition to these innovations were firehouse registers, gongs and bells that struck in rhythm to the number of the alarm box that had been sounded.

Gamewell developed electromechanical devices which, when the alarm struck, automatically caused the chains across the horse stalls to drop. The horses were trained to stand under the harness as firemen quickly made the hitch. From the time the alarm was pulled until it was received in the Gamewell-developed fire alarm offices and was repeated to the firehouses, apparatus was going out the doors in only a minute or so.

All of which is not to suggest that, despite constantly improving fire alarm systems, the occasional glitch sometimes made history. History's most famous occurred in Chicago on the evening of Sunday, October 8, 1871. When Mrs. O'Leary's famous cow kicked over that famous lantern, the resulting fire ignited the barn.

William Lee, a passerby, saw the fire. In his excitement to turn in the alarm, he bypassed several fire alarm boxes and ran three blocks to Twelfth and Canal Streets where Bruno

H. Goll operated a drugstore and kept a key to Box 296 at the corner. Goll unlocked the door of the firebox and turned in the alarm at 9:05 p.m. For some unknown reason, the alarm was never received at the Courthouse fire alarm headquarters.

With the O'Leary barn fire burning for around half an hour and spreading to other barns and shacks, everyone in the neighborhood listened for the approaching clang of the horse-drawn apparatus. When none came, Goll turned in the alarm again. Still no fire apparatus which, historians believe, could have controlled the flames.

The comedy of errors was compounded in the Courthouse alarm office where Fire Alarm Operator William J. Brown was entertaining several women by strumming his guitar. One of the women noticed a glare off to the southwest and called Brown's attention to it. He passed the glow off as simply a rekindle of a large fire the previous night. Watchman Mathias Schaffer in the Courthouse tower also saw the glow and focused his spyglass on it. He, too, thought it was a harmless rekindle of the earlier coal yard fire.

Engine 6, upon returning from the coal yard fire which they had fought all night and that day, went to bed while Firefighter Joseph Lauf stood the first watch in the firehouse tower. He saw the O'Leary flames and turned out the company. In those days before radio communication among fire companies, Engine 6 was alone in answering the alarm.

At 9:21, nearly an hour after the O'Leary fire started, Schaffer looked again. The glow was worse. Putting aside his spyglass, he called down the speaking tube to Brown: "Strike Box 342, Canalport and Halsted." The best that can be said for that action is that an alarm had at last been transmitted. But to a location more than a mile south of O'Leary's barn.

Schaffer focused his glass on the glow again, realized his error, and tried to correct it. Schaffer refused to strike the correct fire alarm box number. "It will only confuse the companies," he said. "They'll see the fire and go to it." And that's exactly what happened.

The firemen did not find the fire; the fire found them. By the time apparatus turned around and headed for the O'Leary's, Chicago was well on its way to a conflagration which burned for more than 30 hours, destroyed one-third of the city and carved a swath more than five miles long and one mile wide. Destroyed were 17,450 homes and hundreds of factories, stores and other buildings. The death toll was estimated at around 300.

Chicago's nightmare was not, of course, Gamewell's fault. It did point up, however, that no matter how failsafe an alarm system may be, there is plenty of room for human error and requirements for backups in the event of system breakdowns. To the credit of Gamewell and his colleagues, plus those in the many other companies which supplied fire alarm equipment, as quickly as a problem arose, they set about solving it.

While nobody realized it at the time, Alexander Graham Bell's invention of the telephone in 1876, not only revolutionized communication in the United States and elsewhere, but was the beginning of an era when fire alarm boxes would become as anachronistic as horses pulling fire engines. Even with the successful development and extension of telephone systems, phone companies, schools and, of course, fire alarm producers emphasized the slogan: "Know the Location of Your Nearest Fire Alarm Box."

A New York City fire alarm box placed in service on March 25, 1870. A key was required to open the outer door. Policemen and nearby storekeepers had keys for opening these types of boxes. This model was located outside New York public schools as late as the 1950s. Steven Scher

The author of this book quickly learned in school that the alarm nearest his Buffalo home was Box 1531, Mapleridge Avenue and Deerfield Street. I never had to pull it, but to this day — 40 years later — I recall exactly which street corner on which it stood.

The only time I used an alarm box was at 12:29 p.m., October 25, 1951, when I pulled Box 8124 at Diversey Avenue and Clark Street for a fire in the basement of a large drugstore on that corner. Three engines, two hook and ladders, a squad and two chiefs quickly answered the alarm. Third Division Marshal Raymond Daley subsequently called for a 2-11 (second) alarm and extra equipment following an explosion which hospitalized 27 firemen and injured as many more.

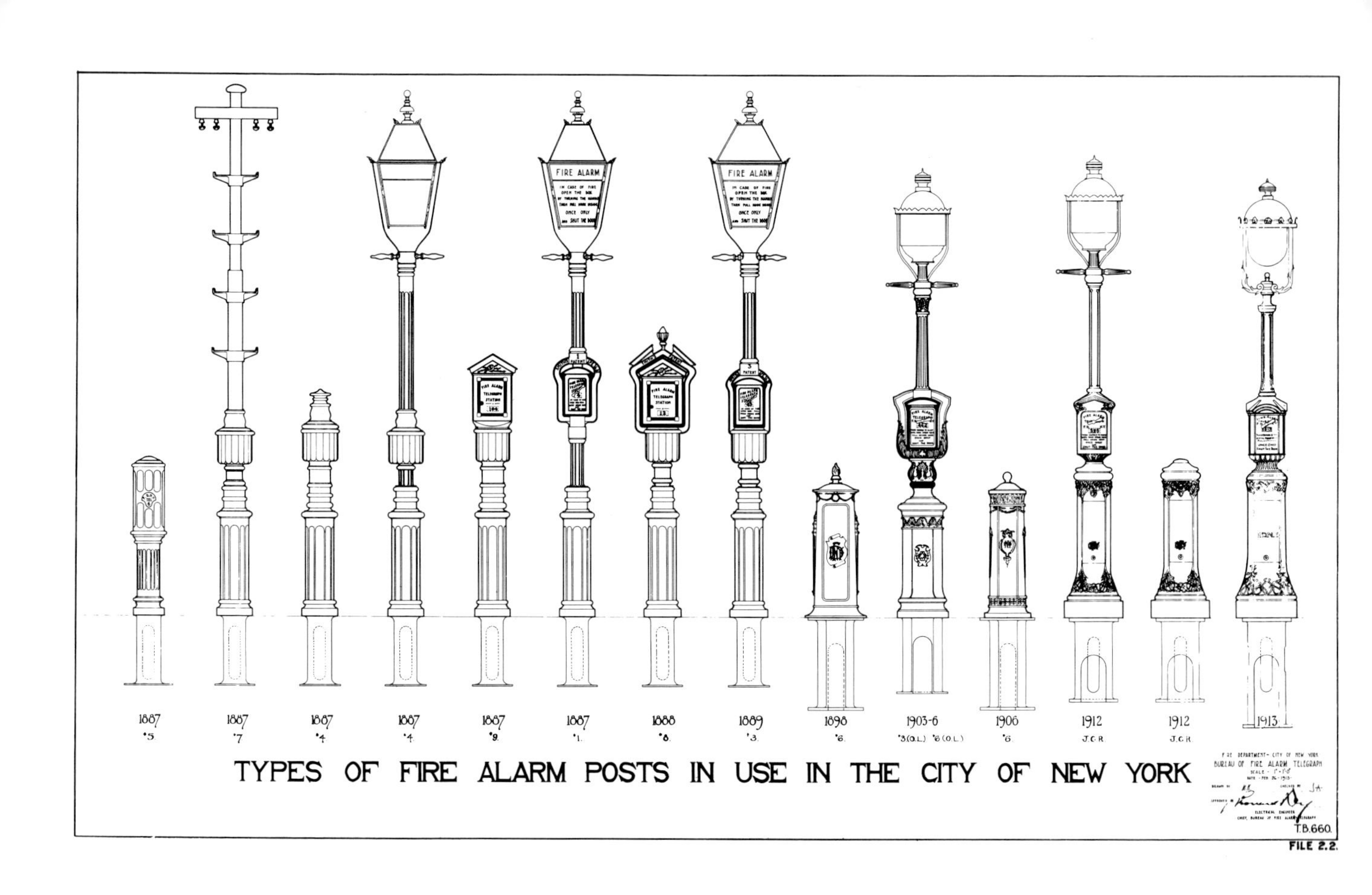

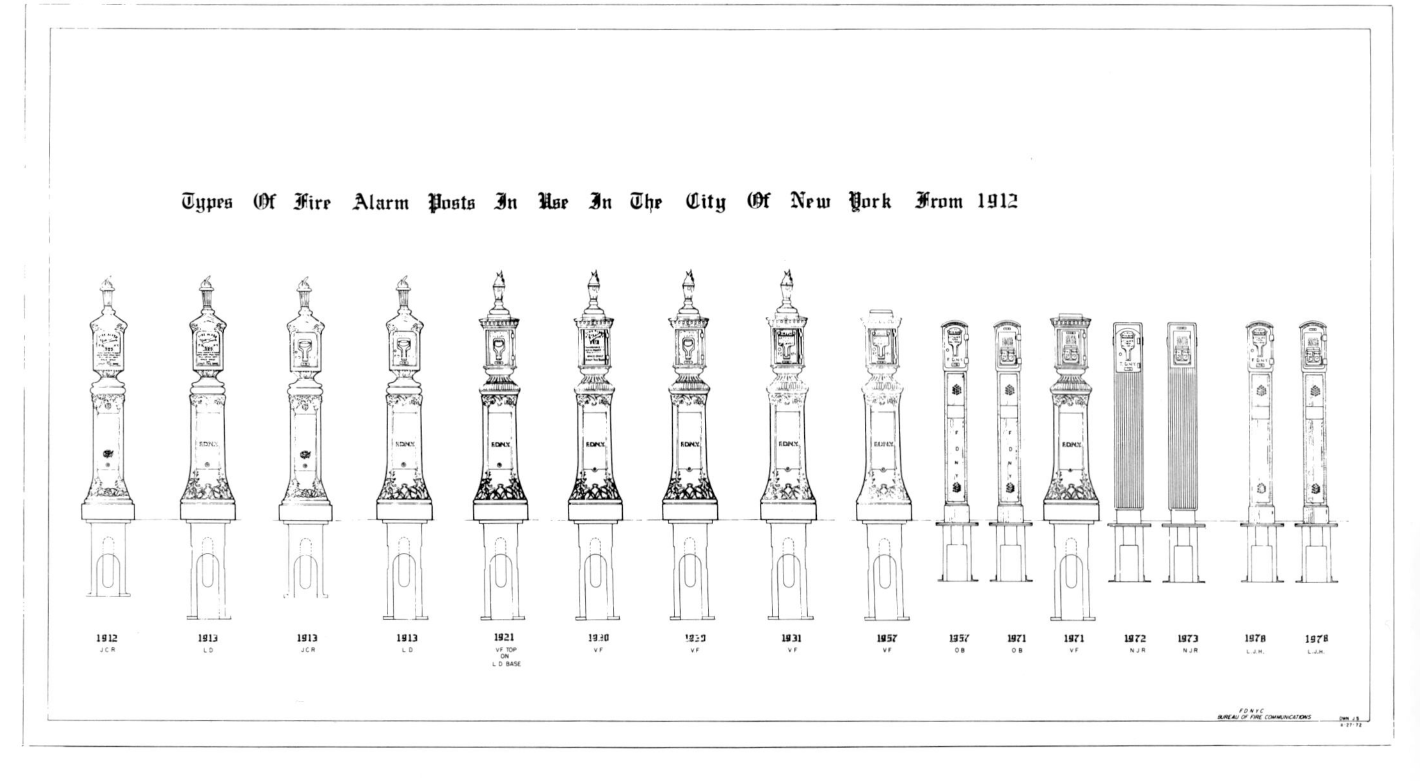

Steven Scher

There were no fire alarm boxes in Los Angeles, July 3, 1980, when I discovered an attic fire in my Woodland Hills home. I telephoned the alarm at 4:17 p.m. Through a misunderstanding both in the fire alarm office and among nearby fire companies, the firemen were greatly delayed. I tried to call again, but the phone lines had burned. When firemen arrived, they called for a second alarm, but by that time flames were causing over $200,000 in damage. A fireman was severely injured while battling the blaze.

A stark comparison between the location of a fire as turned in by telephone and an alarm box occurred in the crowded Our Lady of the Angels Roman Catholic School, shortly after 2 p.m., December 1, 1958, in Chicago. Fire, starting among papers stored at the bottom of an open stairwell, spread smoke and flames throughout the old, unsprinklered two-story brick building.

Smoke and heat made corridors and classrooms untenable, but a nun ran to inform the mother superior. She could not be found, so the teacher pulled the school's fire alarm system which rang, but did not send an alarm to the Chicago Fire Department. Many teachers and children were trapped in their classrooms. A passing motorist saw smoke, but was not able to use a public telephone. The school janitor saw the smoke, ran to the rectory and told the housekeeper to telephone the fire department.

The call, received at the City Hall Main Fire Alarm, was logged at 2:42 p.m., many minutes after the fire started. The housekeeper was almost unintelligible, but dispatchers managed to piece together that there was a fire at 3808 Iowa Street, half a block from the school. Only three pieces of equipment were dispatched, which was standard protocol under the circumstances. Arriving firemen not only found a fierce fire, but children hanging and jumping from windows.

By then, 15 additional telephone calls, mostly giving widely-scattered locations of the fire, flooded into the alarm office. Chief Fire Alarm Operator Joseph Hedderman wisely determined the nearest fire alarm box to the school and transmitted Box 5182, which sent more fire apparatus. By then, most, if not all, the children who perished were already dead due to the delays caused by the telephoned alarms.

Chicago's infamous Our Lady of the Angels fire resulted in a 5-11 (fifth) alarm. A total of 92 children and three teachers perished. The disaster stunned parents and school officials throughout the world. One of the results in Chicago was the installation of a fire alarm box in front of every school. Many other communities followed suit.

Boxes usually are linked to the school's fire alarm system and either automatically transmit the alarm or call firemen when the box is manually pulled. Running card assignments for these school boxes, as well as boxes placed outside hospitals and similar institutions, require a larger than normal assignment of engines, aerial ladders, squads and other equipment. Since that horrendous day in Chicago there has not been a comparable school fire in the United States.

There was, therefore, much to be said in favor of the fire alarm box over the telephone. After Bell's invention, it would be many years before telephones were found in nearly every home. Telephone calls reporting fires often result in transposition of numbers of the address of the fire. Similar excitement, if not panic, often sends fire companies to wrong addresses while the fire itself burns. Still another point favoring alarm boxes was their heavy usage in seaports. A sailor who could not speak English could pull an alarm box hook and there would be no doubt of the fire's location. The same holds true in 1990 as more than a dozen languages are spoken in many large cities.

Los Angeles fireman counts the holes on the tape as a box alarm is received in a firehouse. Directly under the register is a list of boxes and locations to which this company is to respond on the first alarm. Paul Ditzel

But there were drawbacks, too. Persons turning in a box alarm were encouraged to remain at the box until firemen arrived so they could be directed to the fire. When large fires occurred, many boxes were often pulled; some far distances from the actual fire. Perhaps the most famous example occurred shortly after 2:16 a.m., June 22, 1947, when the tanker *Markay* exploded and burst into flames in the Wilmington area of Los Angeles Harbor. The first alarm came from Box 15 in San Pedro, a mile away. Had firemen not heard and seen the explosions and fire, they would have been delayed many minutes in reaching the fire.

Box alarms could be wasteful of fire department resources. A box alarm called a pre-determined number of engines, hook and ladders and other equipment. Perhaps the alarm was for a wastebasket fire, which one engine could handle. Perhaps it was a lumber yard where more help would be needed.

Firemen rarely knew what awaited them when they arrived at the fire alarm box. As years passed, box alarms became increasingly false. Slow police and ambulance response often resulted in boxes being pulled for strictly

THE

Gamewell Fire Alarm Telegraph Co.

No. 1½ BARCLAY ST., NEW YORK.

PROPRIETORS OF

The "Old and Only Reliable" American

Fire Alarm

AND

Police Telegraph.

In successful operation for more than a quarter of a century.

IN USE

BY NEARLY

250 OF THE LEADING CITIES

IN THE

UNITED STATES

AND

CANADAS.

Our systems contain all the most modern improvements, and are furnished at the lowest prices.

Fire Committees should write us for estimates before adopting any other system.

JOSEPH W. STOVER, Boston,
President.

D. H. BATES, New York,
1st Vice-President.

W. H. WOLVERTON,
2d Vice-President.

C. W. CORNELL, New York,
Sec. and Treasurer.

J. N. GAMEWELL, New York, General Superintendent.

E. B. Chandler..............118 La Salle St., Chicago, Ill.
Edwin Rogers...............27 Federal St., Boston, Mass.
Utica Fire Alarm Telegraph Co...........Utica, N. Y.

J. F. Morrison, 32 & 34 E. Monument St., Baltimore, Md.
Webb Chandlee..........................Richmond, Ind.
California Electrical Works, 35 Market St., San Francisco.

SEND FOR CATALOGUE.

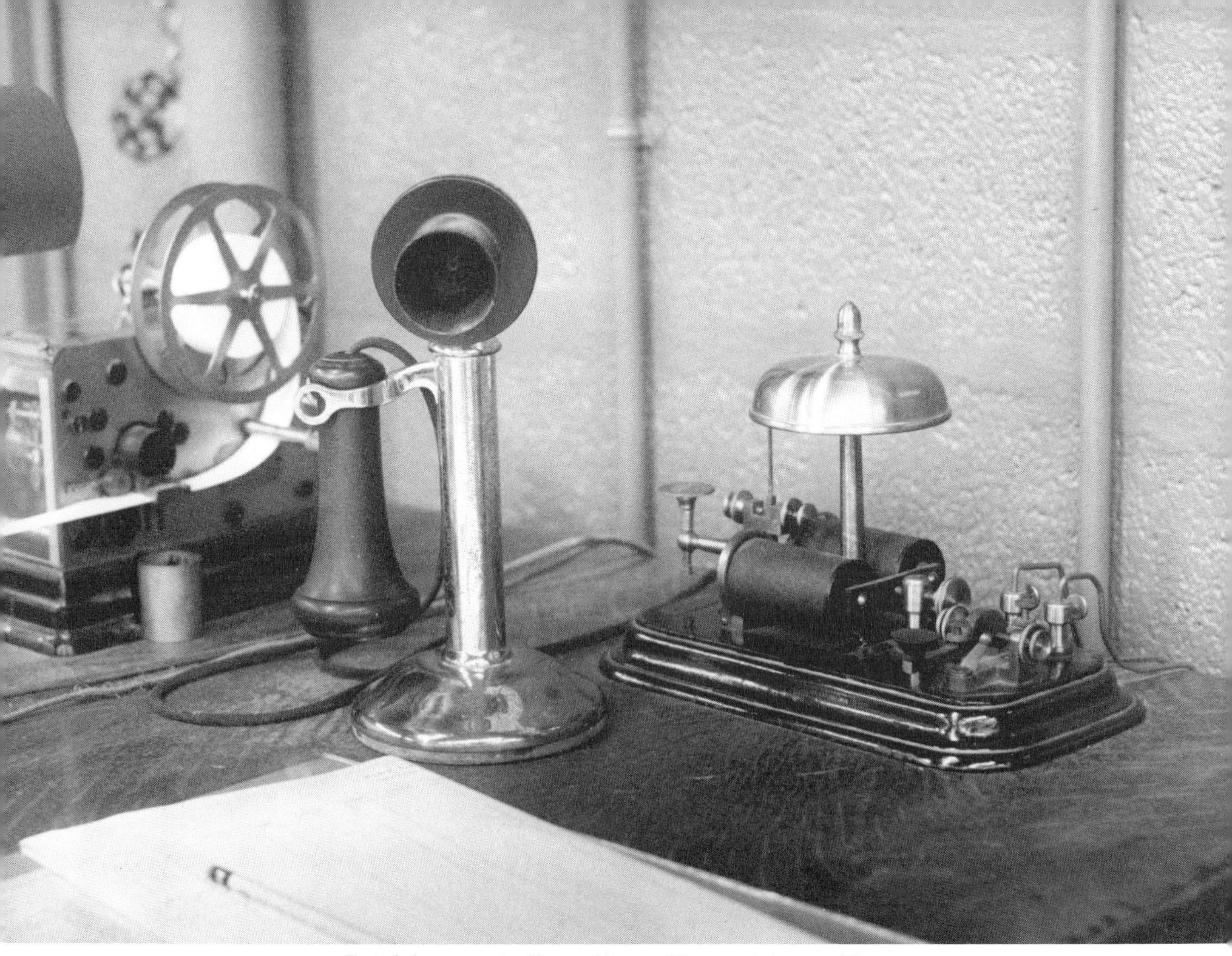

Typical alarm center in offices and homes of fire commissioners and fire chiefs around 1920. The alarm register, with its spool of paper tape is at the left. In the center of the photo is the direct telephone to the central fire alarm office. A combination Umbrella or Turtle tapper bell and telegraph is at the right. Steven Scher

those needs. The public quickly came to realize that it might wait an hour for policemen, but they would get plenty of help fast if they pulled a fire alarm box; even if it was for a minor street scuffle.

An amusing — but not to Buffalo firemen — were false alarms turned in by visitors from nearby Canada. Postal mailboxes in the province were painted red. Occasionally a Canadian visitor, mistaking a Buffalo firebox for a mailbox, would pull the hook and look for the mail drop to open. Instead, he quickly heard the rapid approach of fire engines, hook and ladders and battalion chiefs.

Even more of an inducement for fire alarm box protection was contained in standards set down by fire insurance ratings and the National Board of Fire Underwriters. Starting in 1904, and continuing to the famous NBFU Pamphlet No. 73 in September, 1949, specific standards were established "for the installation, maintenance and use of Municipal Fire Alarm Systems:

In its 39 pages, the National Board covered everything from cable sizing to current protection and supply. Excerpts are of particular interest:

"For communities receiving less than 600 alarms per year, alarms not retransmitted automatically shall be received and retransmitted to the fire force by a responsible person on duty...."

"For communities receiving more than 600 and less than 1500 alarms per year, at least one operator especially trained for the service shall be on duty at all times. Where a street box signaling system is maintained, the operator shall be in the central fire alarm office and be capable of testing and operating the system.

"For communities receiving more than 1500 alarms per year, at least two fully trained and competent operators shall be on duty at all times.

"A Type A (manual) system is one where alarms from street boxes require an operator to check their receipt and to retransmit all alarms over two classes of alarm circuits to fire stations. Facilities for automatic retransmission of alarms from street boxes may be provided and may be normally in service except where the number of alarms exceeds 1500 per year....

"A Type B system (automatic) is one where alarms from

Shoulder patch worn by members of the alarm dispatchers and operators in the Fire Department of New York. Steven Scher

Fire Alarm Operator Mat Tomlan of the Chicago Fire Department checks the red ink dots stamped on paper tape streaming from a fire alarm box register. Ken Little

street boxes are automatically retransmitted over box circuits and usually over a single class of alarm circuits. Type B systems are designed for communities having an organized fire department but not requiring a Type A system and having less than 1500 alarms per year, however received....

"Boxes shall be non-interfering and succession, except as follows:

"In a system requiring not to exceed 20 boxes, the boxes need not be succession, but shall be non-interfering if there is more than one box....

"In a Type A system which is not equipped with automatic repeating means, circuits including only boxes of plain interfering type may be provided, if all of such boxes have closed-type break-wheels, not more than ten boxes are placed on any such circuit, and are so placed that the distance between any two boxes in any one circuit, via highway, will not exceed 1200 feet.

"Non-interference devices either mechanical or electrical, shall be designed so that manipulation of box starting levers, singly, concurrently or consecutively, will not under any circumstances result in a false signal.

"Succession devices, either mechanical or electrical, shall be designed so that no signal will be lost if the starting levers of two boxes are pulled at or about the same time.

"Boxes shall send three or four rounds of box numbers. Four rounds are required where outside alarm devices are operated directly from the boxes for summoning firemen.

"Boxes shall be capable of operating properly at any speed up to 4 strokes per second. For a Type A system they shall be set to operate at not slower than 2 strokes per second. For a Type B system they shall be set to operate at the speed of the slowest instrument connected for response to the circuit.

"Boxes shall have doors which the starting handle is in plain sight and is readily accessible by breaking a glass or opening a cover.

"Boxes shall be located so as to suitably protect the city or town.

"NOTE: In general it is considered that a box should be plainly visible from the main entrance of any building in congested districts. In mercantile or manufacturing districts it should not be necessary to traverse more than one block or more than 300 feet to reach a box; in residential districts this distance should not exceed one block or 500 feet.

"Boxes shall be conspicuously located, at street corners where practicable. The box and a portion of the supporting pole or post shall be painted "Signal" red, preferably with white stripes above and below the red. A special colored light shall be provided at or near every box in closely built sections, to indicate location at night.

"Box numbers should be assigned, as far as practicable, so that consecutive numbers will be closely grouped, and so that the same initial digits will be confined to well defined districts or contiguous territory."

In the standard grading system, points could be deducted for inadequacies in alarm systems. The result, as far as citizens were concerned, could be higher fire insurance premiums.

The second event which ultimately would encourage communities to remove fire alarm systems was the first use of radio in Boston, starting in 1924. Radio was a quicker means of communicating with fire alarm headquarters.

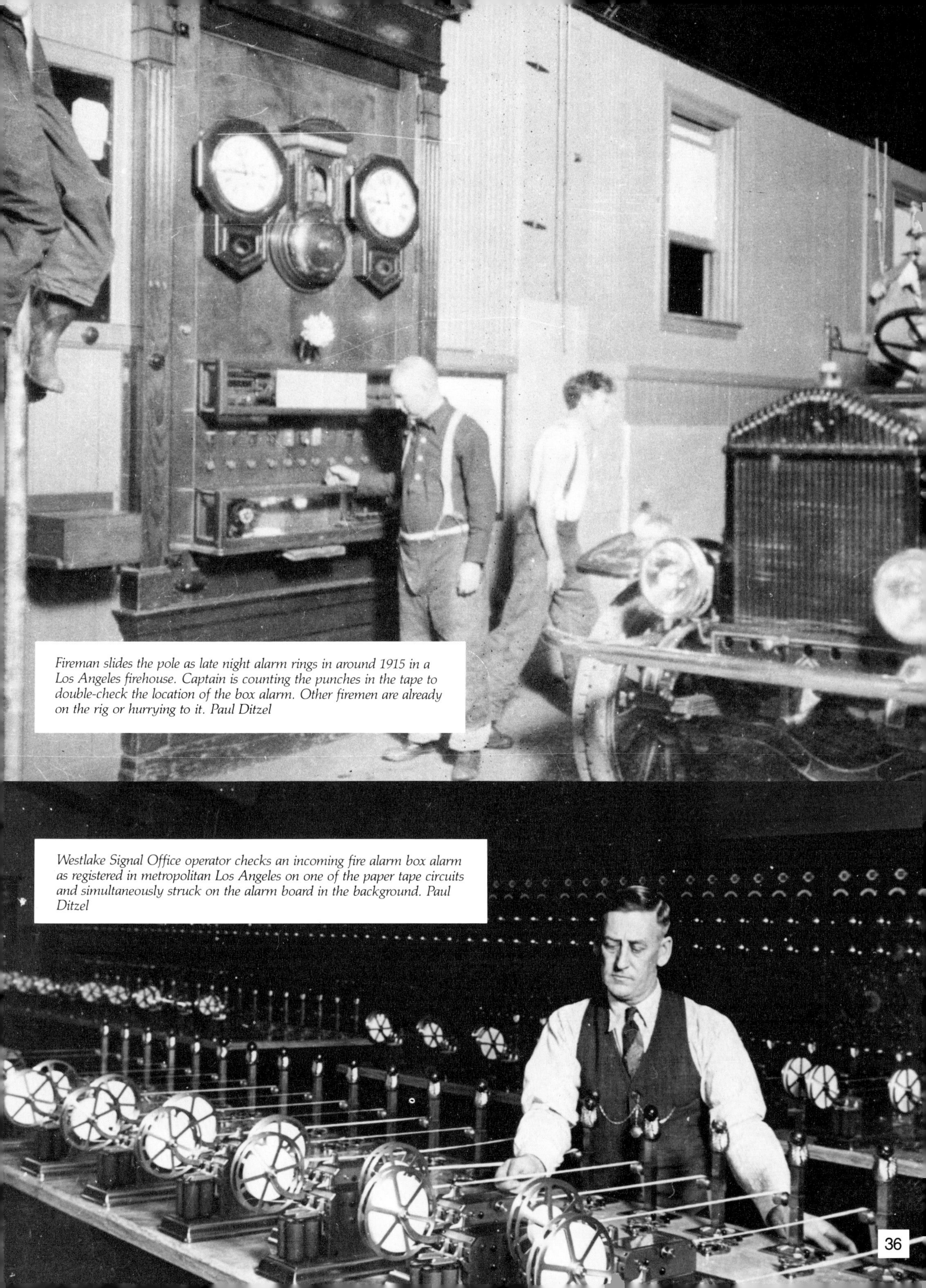

Fireman slides the pole as late night alarm rings in around 1915 in a Los Angeles firehouse. Captain is counting the punches in the tape to double-check the location of the box alarm. Other firemen are already on the rig or hurrying to it. Paul Ditzel

Westlake Signal Office operator checks an incoming fire alarm box alarm as registered in metropolitan Los Angeles on one of the paper tape circuits and simultaneously struck on the alarm board in the background. Paul Ditzel

Prior to its spread throughout the fire service, a fireman, usually the chief's driver, had to go to the nearest fire alarm box upon arriving at a fire. By means of a telegraph key and later a telephone jack, he could communicate with alarm headquarters. He relayed the chief's declaration that the fire was out and no more help was needed, or he could request a second, third, fourth or fifth alarm which would summon additional companies.

The fire service was not altogether quick to accept radio. Firemen could not see the radio transmissions, but they could hear the telegraph clicks inside the box as they tapped messages and received messages from alarm headquarters. Evidence of this skepticism was found in Buffalo, New York. A chief requesting a second alarm would call for it by using the police radio in his car. To make certain the request was received, his driver would immediately go to the nearest fire alarm box and verify, by telegraph key or phone, the chief's order. This system prevailed until around 1950.

With telephones as commonplace as televisions, the days when the little red firebox on the corner was as much a bit of Americana as Cracker Jack boxes were on the wane. The first nudge toward oblivion was the growing false alarm problem, particularly after riots in cities starting with that in the Watts area of Los Angeles in 1965. False alarms had been a mounting problem in all cities. Rioting in Los Angeles and elsewhere only escalated the problem. Alarms often severely degraded protection in high fire incidence areas as firemen answered false alarms from fire boxes.

Reluctant to break a tradition that was more than a century old and sensitive to the possibility of fire insurance penalties, local government officials tried many measures to cope with the problem: sending one engine to a box alarm to investigate. If there was a fire they could call for more help. Invariably the alarm was false.

Some boxes became daily, even hourly sources of false alarms, especially as schools let out for the day. Buffalo, for example, had its famous false alarm Box 936, Virginia Avenue and 10th Streets, whose alarms were false 90 percent of the time. Someone, possibly a neighbor weary of hearing fire apparatus day and night, stole the box from its pole.

Firefighting was, under the best of conditions, a highly hazardous occupation. Fire service history contains many examples of firemen being killed in crashes while answering false alarms. Some cities addressed the problem by installing a device which handcuffed the person turning in the alarm. Others developed small booths from which the person could not leave after sending an alarm. From handcuffs or booths, release came only when firemen arrived to free them. Philadelphia equipped boxes notorious for false alarms with an air horn that could be heard for blocks around when the box was pulled. None of these systems worked.

Fortunately, Gamewell never knew of the time when firemen came to regard a firebox alarm as a false alarm. His later years were occupied with expanding the Gamewell name and acceptance. If American LaFrance was the General Motors of fire apparatus, Gamewell enjoyed a similar renown and comparison in its line of equipment.

Gamewell's later years were marked by steady bickering over claims to credit for the successful fire alarm systems bearing the famous lightning rod Gamewell logo that was cast into the front of the fireboxes. These problems, apparently, were blown far out of proportion by relatives of Moses G. Crane, who rightfully deserves much of the credit for Gamewell's success.

Crane, the very stereotype of the introverted clockmaker, devised many technological breakthroughs which Gamewell proceeded to sell. In retrospect, it seems academic that Crane's family attempted to goad him into claiming more credit for the success of Gamewell. Evidence indicates Crane was content to devise and invent, while letting the glory go to Gamewell.

Suffering from advanced tuberculosis, the ills of old age, and involved in stressful patent lawsuits, Crane took a pistol on July 7, 1898, and put a bullet through his head.

Gamewell preceded him in death by two years. Suffering intensely from heart problems and infirmities of age, John Nelson Gamewell, 73, died in his Hackensack residence at 3:30 a.m., Sunday, July 19, 1896. His death was little noted outside of New Jersey. But the Gamewell name was to endure as America approached the 21st Century. If Gamewell's fire alarm system has become outmoded, the name Gamewell will forever be synonymous with the very finest in fire alarm technology so long as one of his little red fireboxes stands on any street corner in any American or overseas community.

Ken Little (standing, left) and Francis Bodkin (standing, right) operating Main Fire Alarm Office of the Chicago Fire Department, along with John Hankins, Martin Paulson and Dan Mahoney at the console. Ken Little

Famous display of all Gamewell equipment in the company's Exhibition Room at its offices in New York City.

Paul Ditzel

CHAPTER FOUR

FIRE ALARM ALMANAC

*1788 — **MARCH 21.** Good Friday conflagration in New Orleans. Priests forbid ringing of church bells to sound fire alarm because it was a Holy Day. They said the fire was God's punishment for sins of the citizens.*

*1844 — **MAY 24.** Samuel Morse demonstrates his telegraph to Congressmen by sending the message, "What hath God wrought," over lines strung from Washington to Baltimore.*

*1845 — **JUNE 3.** Dr. William Francis Channing first describes his fire alarm box system to the Boston Daily Advertiser.*

***1845.** First firebell-watchtowers built in New York. Other cities relied on church and government-owned building bells to sound fire alarms. Smaller communities used gunfire, fox horns, iron hoops, triangles, wood rattles and other devices.*

***1848.** Moses G. Farmer invents a bell-striking device for sounding fire alarms.*

***1850.** Charles Robinson, in New York, is the first to use Morse telegraph a transmit fire alarms.*

***1851.** New York connects its eight firebell-watchtowers with telegraph lines.*

***1851.** Boston appropriates $10,000 for construction of Channing and Farmer's fire alarm system.*

*1852 — **APRIL 28.** Channing and Farmer fire alarm system completed. The original Boston box is made of metal and wood. Inside the box a telegraph key was opened and closed by the turning of a crank.*

*1852 — **APRIL 29, 8:25 P.M.** First fire alarm sounded from a firebox.*

***1854.** Channing and Farmer apply for a patent for their fire alarm system.*

***1855.** Channing describes his Boston fire alarm system at the Smithsonian Institution. John Nelson Gamewell is in the audience.*

***1855.** Gamewell acquires regional rights to market Channing and Farmer's fire alarm system. Four years later he purchases all rights for around $30,000.*

*1855 — **1861.** Gamewell builds five alarm systems around the United States: Philadelphia, Baltimore, St. Louis, Charleston, S.C., and New Orleans. The Civil War as General William Tecumseh Sherman invades South Carolina.*

*1861 — **1865.** United States Government confiscates Gamewell's patents during the Civil War as General William Tecumseh Sherman invades South Carolina.*

***1861.** Boston changes color of fire alarm boxes from black to red, which was to become the universally recognized color of boxes throughout America.*

***1865.** John F. Kennard of Boston buys Gamewell's confiscated patents for $80. Forms partnership with Gamewell.*

***1865 (circa).** Installation of firehouse gongs results in box numbers being transmitted directly to the stations.*

***1867.** Spring-driven boxes begin to replace those driven by weights.*

***1868.** Moses G. Crane patents his first electromechanical gong.*

Automatic whistle-blowing apparatus.

1869. Crane and Edwin Rogers patent a non-interference signal box: If two or more boxes are operated at or about the same time, one box will be automatically selected to control the circuit and send a correct signal; excluding from operation the other box or boxes simultaneously operated. Crane and Rogers replaced older weight-driven mechanisms in boxes with a spring-driven devices. Pulling a lever released the clock-like mechanism that turned the code wheel containing teeth or ratchets according to the number of the box.

1869 — AUGUST 1. Charles T. and J. N. Chester contract with New York City for a fire alarm box and central office system. The Chesters' company was later acquired by Gamewell.

1870. Rogers patents the first automatic repeater for receiving and retransmitting fire alarms from boxes.

1871. Gamewell patents a distance non-interference box which avoided, but did not always prevent, interference between two (or more) boxes when pulled. Not until 1875 was the problem properly addressed by a system devised by James M. Gardiner, Gamewell's brother-in-law.

1871 — APRIL 7. Gamewell patents a modified code wheel which operated when the electric circuit was opened instead of closed.

1871 — OCTOBER 8. Great Chicago fire largely results because fire alarm box failed to operate. Fire alarm operators further blundered when they failed to immediately strike a box alarm. When the box was finally struck it was distant from the actual fire. By the time the full resources of the Chicago Fire Department managed to make their way to the fire, the entire neighborhood was in flames and the fire was beyond control.

1873. Cincinnati Exposition. Sir William Thompson and other exposition judges cite the Gamewell system as "the most ingenious and finest piece of telegraphic mechanism ever exhibited."

1873. J. M. Fairchild develops a trap lock for fire alarm boxes. A person using his key to turn in an alarm would find that the trap mechanism would not release the key. This was one of the first breakthroughs in the prevention of false alarms.

1874. Henry Parmalee invented the first practical automatic fire sprinkler heads and installed them, along with piped water supplies, throughout his Mathushek Piano Company factory in New Haven, Conn.

1874. American District Telegraph Company is formed in Baltimore. ADT was to become an industry leader in developing sprinkler and other alarm systems. A fire, setting off one or more sprinkler heads, automatically

Central Fire Alarm Office of the San Francisco Fire Department. Willy Dunn

transmitted a coded alarm to a central office and sometimes directly to fire alarm offices and firehouses. ADT watchman could also pull ADT fire alarm boxes in subscribers' buildings. A water flow alarm or a manual pull quickly brought firemen. Sometimes these boxes were connected to city fire alarm boxes which automatically relayed the alarm to fire alarm offices or stations.

1875 — JUNE 15. *Charles Tooker of Chicago patents a keyless firebox door. The Tooker door could be opened by turning a handle. When opened, the door caused a bell in the box to ring. To complete the alarm, the inside lever had to be pulled. Although the ringing bell alerted those nearby that the box had been opened, it was not altogether clear that the lever had to be pulled to sound the alarm. This somewhat confusing system, especially in the excitement caused by a fire, resulted in many false alarms.*

1875. *Gardiner patents (for Gamewell) an improved non-interfering box. (See 1869 and 1871.)*

1876. *Alexander Graham Bell invents the first practical telephone. Not only did it provide a new means for notifying firemen of emergencies, but it was to be freely-used in fireboxes and for communications among alarm offices and firehouses.*

1876. *Joker System invented and developed. The Joker was usually mounted on or near the firehouse watch desk and provided telegraph and telephone communication with the main fire alarm office, or offices, as well as other fire stations. Instruments included an alarm register, a small gong, telegraph sounder, polarized relay and a telephone. An advantage of the Joker was that the telephone or telegraph key could be operated without interrupting a fire alarm signal being received on the register, tape, bell and paper tape take-up reel.*

1879. *Gamewell forms the Gamewell Fire Alarm Telegraph Company. Incorporation papers show John N. Gamewell, proprietor, American Fire Alarm Telegraph.*

1879. *Thomas Alva Edison invents the first practical electric light bulb. Within a few years, light bulbs and lanterns were used to clearly identify fireboxes at night.*

1880. *Gamewell introduces the Excelsior gong with a 6-inch-diameter bell.*

1880 (circa). *Fire Department of New York starts conversion to Gamewell system.*

1880 (circa). *Gamewell introduces the Excelsior Box and the No. 4 Sector Box. The Excelsior Box, similar to the Gardiner Box, originally was an interfering type of box, but subsequently was changed to the non-interfering versions. (See Cavalcade of Fire Alarm Boxes.)*

1881. *Development in Providence, Rhode Island, of the first auxiliary fire alarm system known as Rogers. These boxes, located inside buildings, did not have code wheels. When actuated, usually by a small switch, they sent an electrical impulse to the nearest street box which automatically transmitted the alarm to the fire department. San Francisco claims credit for the first extensive auxiliary system.*

1886 — OCTOBER 5. *Crane sells his business interests to Gamewell for $4,200. Five years later, he again entered the business, but did not do well.*

1885 — OCTOBER 27. *Lewis H. McCullough patents central office apparatus for fire alarm telegraphs. The system, as developed, is an arrangement of devices and wiring which provides a method for obtaining signals over fire alarm circuits in the event of open circuits or electrical grounding. The system can be automatic or manual in operation.*

1886. *Gamewell systems are installed in 250 cities, mostly in the United States with a scattering in Canada. Four years later, Gamewell systems number 500. By 1910, Gamewell held a 95 percent market share, despite competition, some of it stiff, from about three dozen other manufacturers.*

1889. *John J. Ruddick develops the Successive Box mechanism. Prior to the Ruddick mechanism, a box that was pulled on an already busy circuit would be prevented from transmitting an alarm by whatever other box (or boxes) were initially activated. Authorities, acknowledging that this second box pulled could be for another fire (and sometimes was) found an answer*

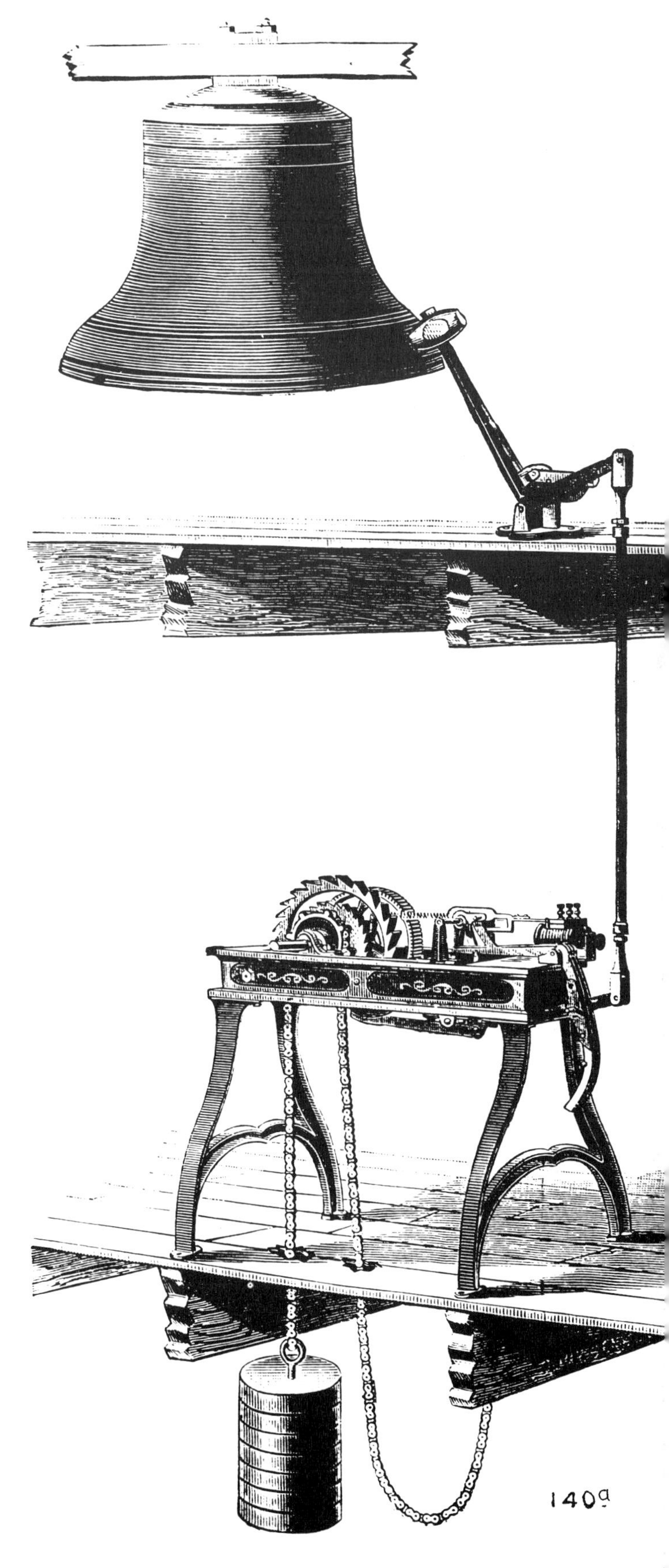

Automatic bell-striker

Introduced by the Los Angeles Fire Department on December 24, 1932, and perhaps unique in the American fire service, was this box alarm transmitter equipped with the Hold-Out System. The operator inserted a perforated card showing the number and location of the box. The perforations, electrically-read, caused large bells to ring only in those fire stations which were to answer the alarm. Other stations were simultaneously notified to standby for possible additional alarms as the system caused small, secondary, bells to ring. Paul Ditzel

in the Ruddick mechanism. With the Ruddick system, the mechanism would run at idle for a short period and immediately send its alarm as soon as the first box ended its transmission. Subsequent improvements led to Gamewell's Three-Fold Box which sent a signal at the end of its 24th round, thus interrupting the circuit if it was still busy.

***1890 (circa).** Gamewell opens a new factory in Newton Upper Falls, Massachusetts, to meet the phenomenal demand for his products. A three-story brick addition was added in 1903 and the complex was further enlarged in 1914. During the 1890s, Gamewell owned more than 200 patents; mostly for technologies invented and designed by others, including clockmakers and fire department alarm operators. Many of them became Gamewell employees. Contemporary accounts describe his monthly payroll totaling $1,000. That figure does not truly represent the number of his employees. Notoriously penurious, Gamewell paid low dollar for highly-skilled talent — both engineering, manufacturing and sales — while becoming wealthy as his name and his name alone, became synonymous with fire and police alarm signaling systems. Gamewell, himself, mostly operated out of his New York City offices which featured a full display of all Gamewell equipment.*

***1895.** N. H. Suren invents a self-starting or automatic door which solved the major objection to the Tooker Door (see 1875). Features of the Suren door included faster transmission of alarms with no room for confusion by the ringing of the bell in the Tooker-type box. Turning of the Suren handle sent the alarm. The alerting bell did not ring until the code wheel was operating.*

***1896 — JULY 19.** John Gamewell, founder of The Gamewell Fire Alarm Telegraph Company, dies while sleeping in his Hackensack, New Jersey home. He was 73.*

***1898 — JULY 7.** Moses G. Crane dies of a self-inflicted pistol shot through his head.*

__1901 — MARCH 19.__ Channing, the father of the fire alarm box system, dies. Shortly before his death he visited Boston's fire alarm office where he marveled at the technological progress that had been made since his first system in 1852. "Well, the baby has grown!" he said.

__1902.__ C. E. Beach makes significant improvements upon alarm repeaters earlier devised by Rogers and Crane.

__1904.__ Frederick W. Cole, a Gamewell engineer, is issued a patent for a Turtle Gong for firehouse use. The gong, made in 6-inch and 10-inch bell dome size, derived its name from its turtle-like appearance. Turtle Gongs offered many improvements over older bells, including fewer moving parts and adjustments, plus faster operation: two blows per second.

__1906 — APRIL 17.__ A major earthquake and resulting fires devastates San Francisco. Severe jolts knocked alarm office batteries off their racks. With lines down and broken, the fire alarm system was virtually useless and was a contributing factor to the catastrophic losses.

__1906 — JUNE 26.__ Cole is issued a patent for his alarm box key guard. Familiarly known as the Cole Key Guard, the device enclosed the key to the outer door of the box within a glass-windowed enclosure which projected out from the box. This prevented its being frozen and inoperable. This type box became standard until the advent of the Quick Action Door in 1922.

__1916.__ Gamewell introduces its Peerless Positive Non-Interfering Successive Fire Alarm. This box, which virtually set the standard by which all future boxes would be manufactured, improved upon earlier models, notably Gardiner and Ideal boxes. (See 1875 and Cavalcade of Fire Alarm Boxes.) Offered in several models, including the standard cottage-shape as well as an oval-shaped box, Peerless boxes attempted to enter the circuit with each rotation of the code wheel. Cole, in describing the Peerless line said, "This box embodies the highest type of pull wind or sector form of signaling mechanism, protected in the best manner to prevent injury of any kind to the box itself or to any person from high potential currents. The best forms of non-interference principles and devices not only prevent interference between its own and other signals, but prevent interference with its own operation from repeated operation of the pull or winding handle."

__1922.__ Gamewell offers an optional telephone jack in its boxes for two-way voice communication between chiefs at a fire and the fire alarm office.

__1922 — JUNE 20.__ Fire Department of New York opens its Manhattan Fire Alarm office in Central Park. The system included 1528 alarm boxes, 212 circuits and was to become one of the world's busiest and most famous alarm offices.

__1922 — DECEMBER 27.__ George Alfred Jackson applies for a patent for Gamewell's first Quick Action Door for alarm boxes. For the first time this gave easier access to the lever that triggered a fire alarm from the box. (See Cavalcade of Fire Alarm Boxes.)

__1924.__ Gamewell improves upon the Quick Action Door mechanism by replacing the traditional small, flat glass covering plate with a convex eyeball-like device which came to be called a Bullseye. Whoever turned in the alarm could see the triggering lever through the Bullseye. (See Cavalcade of Fire Alarm Boxes.)

__1925 — DECEMBER 27.__ Boston opens its state of the art Fire Alarm Headquarters at 59 The Fenway.

__1928.__ Gamewell introduces an aluminum-alloy, Herculite, box. With inner cases of boxes made of Bakelite, a plastic material, the weight of alarm boxes dropped from around 75 pounds to 25. The reduction in weight made boxes easier to remove for maintenance, repairs and return to their posts and pedestals.

__1931.__ Gamewell improves its Peerless line of boxes (see 1916) by adding features, including a McCullough automatic grounding feature. (See 1875). This box, with other advancements (many cosmetic) was known as the first Three-Fold fire alarm Box. (See Cavalcade of Fire Alarm Boxes.)

__1931.__ Gamewell adds white striping to the outer shell of the fire alarm boxes for quicker identification.

Manual transmitter

Smith Keyguard

Gong/Indicator combined

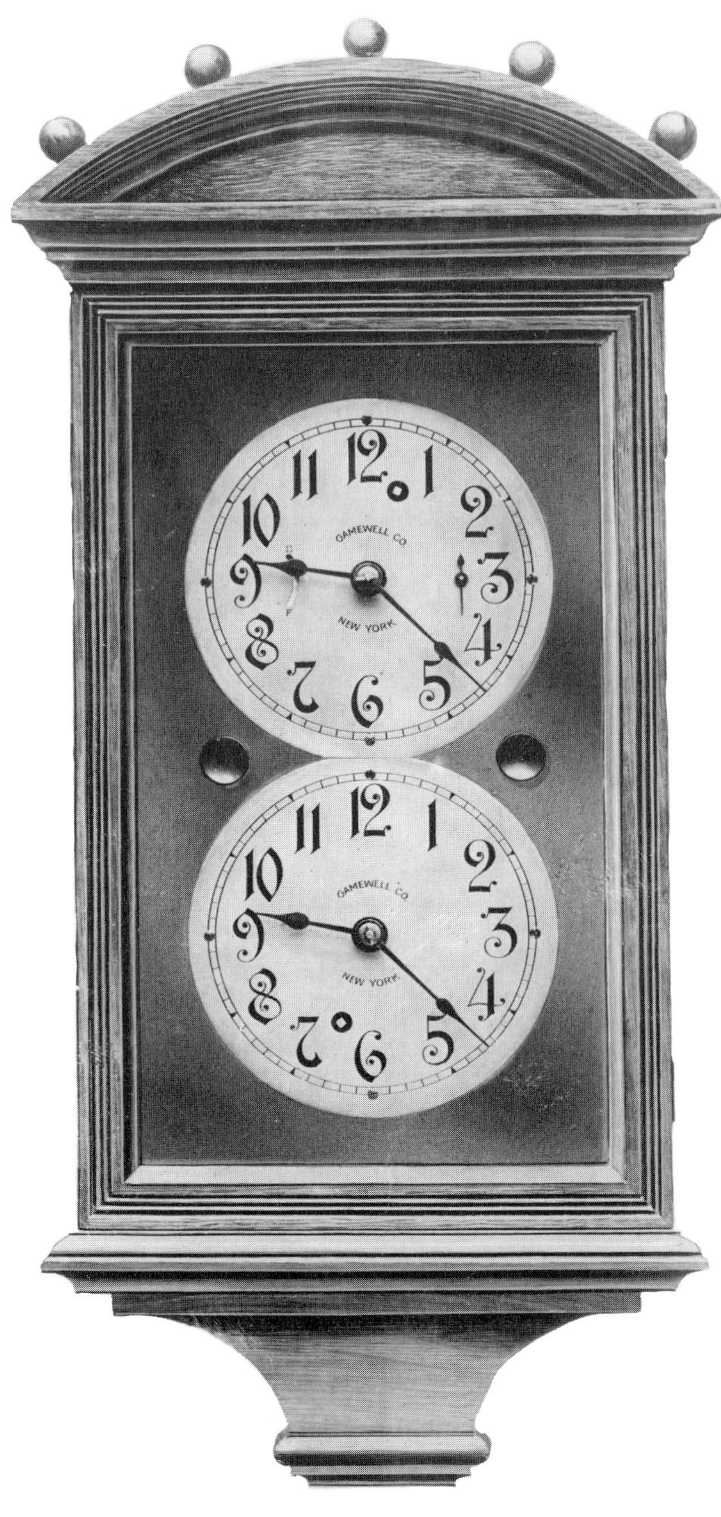

Station clock. One of the dials stops automatically when an alarm is received, thus recording the time of the alarm.

Two early Gamewell registers

Automatic Alarm Repeater housed in glass case.

Early Gamewell House Gong

1932 — DECEMBER 24. *Los Angeles, perhaps uniquely in the fire service, introduces the Hold-Out System. Traditionally, all fire alarms clanged on fire station gongs, whether or not that station was due to answer the alarm. By inserting a perforated card into the Westlake Signal Office alarm transmitter, large gongs rang only in those stations summoned to the box alarm. Other fire stations received the alarm on smaller, secondary, bells to alert them to the possibility that they might be called if more help was needed. This feature was welcomed by firefighters heretofore awakened at night by gongs sounding every alarm box in the sprawling city. A peripheral benefit was the fact that if the large bells clanged, the firemen knew they were being called to an alarm. Valuable time was saved because the ringing of the large bells left no question that the firemen were being called. Until the Hold-Out System, firemen did not know if the alarm was for them until at least one round of the box was received.*

1936. *Gamewell introduced the Vitaguard Fire Alarm System especially designed for communities protected by volunteer fire departments and whose alarm requirements could be served by a system with one firebox circuit which could, if necessary, be divided into two sections or loop circuits.*

1939. *Gamewell replaced its painted white striping on boxes with decals of the word, "FIRE."*

1946 — NOVEMBER 14. *U. S. Attorney General Tom Clark announces Federal Grand Jury indictments of Gamewell, five officers, several affiliates, and The American District Telegraph Company. The indictments charged anti-trust violations in monopolizing the manufacture and distribution of fire alarm equipment. The attorney general said the case "is one of the first to be filed in the program of the Department of Justice to bring relief to public agencies that have been forced to pay exorbitant prices as a result of collusive dealings and monopolistic practices on the part of bidders."*

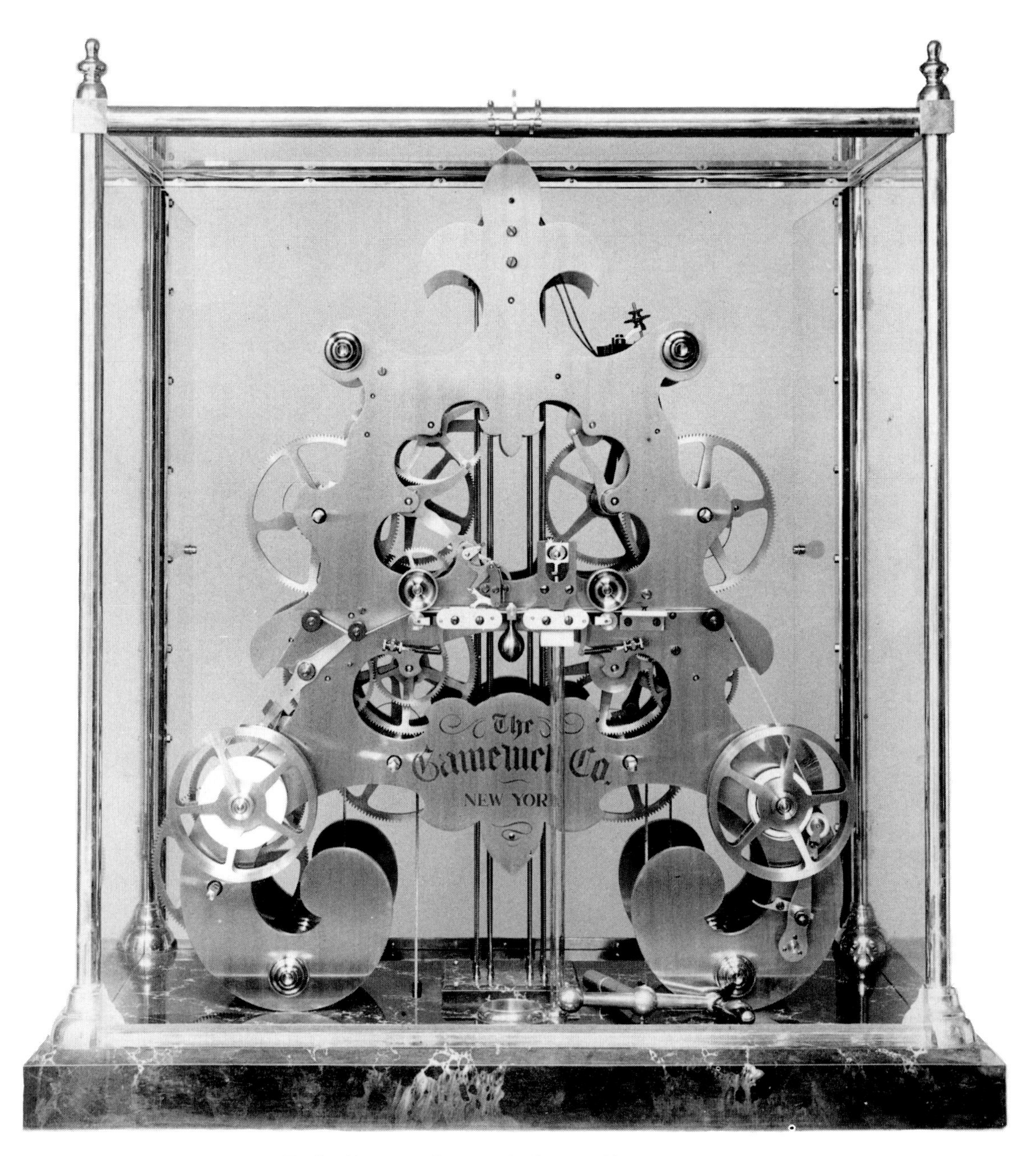

Early Gamewell central alarm office transmitter

1948 — MARCH 22. *Attorney General Clark announced that the defendants, Gamewell and American District Telegraph, had pleaded nolo contendere (no contest) to the 1946 allegations. Fines of $43,250 were levied in the criminal antitrust and the companion civil antitrust case. Gamewell was enjoined from "continuing other of its past practices, such as defraying entertainment costs and other expenses of persons connected with buying such (fire and police alarm) equipment for public bodies."*

1950 — SEPTEMBER. *Dwight G. W. Hollister, board chairman of Gamewell and its subsidiary companies, reports in his 1951 Annual Report that the Justice Department had instituted proceedings in the Federal District Court in Boston. The government alleged Gamewell and "its then president and general sales manager" had violated provisions of the final judgment which had been entered by consent in the company's antitrust cases in 1948, and under which the company has been operating in the public fire alarm field.*

1951. *Hollister, in his Annual Report for this fiscal year, said Gamewell had been fined $50,000, plus costs, and that two company officers, the president and general sales manager, had been given suspended sentences of one year and one day and placed on probation for two years. "In April, 1951," reported Hollister, "the Company filed with the court evidence, which the court accepted as satisfactory, of steps taken to assure future compliance with the consent judgment." Despite Gamewell's problems with the government, the 1951 Annual Report is indicative of how well the fire and police alarm telegraph business was growing. This was a multi-million-dollar enterprise for some three dozen companies fiercely competing in a crowded field. Gamewell nevertheless showed a net income after taxes of $1,165,265.33; compared to $1,259,618.96 for the comparable fiscal year. The publicly-held company paid dividends of $1.50 per share. Even more optimistic times were ahead as the company reported unfilled orders as of May 31, 1951 (the end of the fiscal year) amounting to $10,990,970, a husky increase of $7,242,720 over the prior reporting period. Assets of $9,062,200.38 compared to liabilities of $3,179,075.72.*

1951. *Debate among public officials, telephone and fire alarm officials escalates over the relative merits and drawbacks of alarm boxes and telephones.*

1951. *Gamewell introduces a cosmetic change to the outer appearance of its traditional peaked cottage-shaped boxes. Hereafter, box outer shells are rounded at the upper corners.*

1953. *Miami is the first, or among the first, cities to scrap its traditional fire alarm boxes. They were replaced by bright red (fire) and green (police) boxes with telephones. Phone lines led directly to the emergency communications room. Instead of a traditional box alarm, which could call for around half a dozen pieces of apparatus, the caller could describe the exact location of the incident and the nature of it. Instead of many engines answering a rubbish fire, only one needed to be sent. Similarly, sufficient police resources could be dispatched, depending upon the description of the incident. Miami's system was fiercely debated throughout the emergency services community. Miami said that these street boxes had become preferred over telephones for reporting fires and police calls. False fire alarms became negligible in Miami. The story of telephone alarm boxes remained controversial. Los Angeles, for example, tried them in the harbor area of the city and found that vandals would remove telephones from their cradles as a way of transmitting a false alarm. The telephone boxes were scrapped in Los Angeles. At best, telephone boxes are no panacea for the false alarm problem which continues into the 1990s.*

Upon receipt of an alarm, this device automatically dropped the chains from the horse stalls, allowing the well-trained horses to move forward into position underneath the harnesses.

Early Gamewell House Gongs

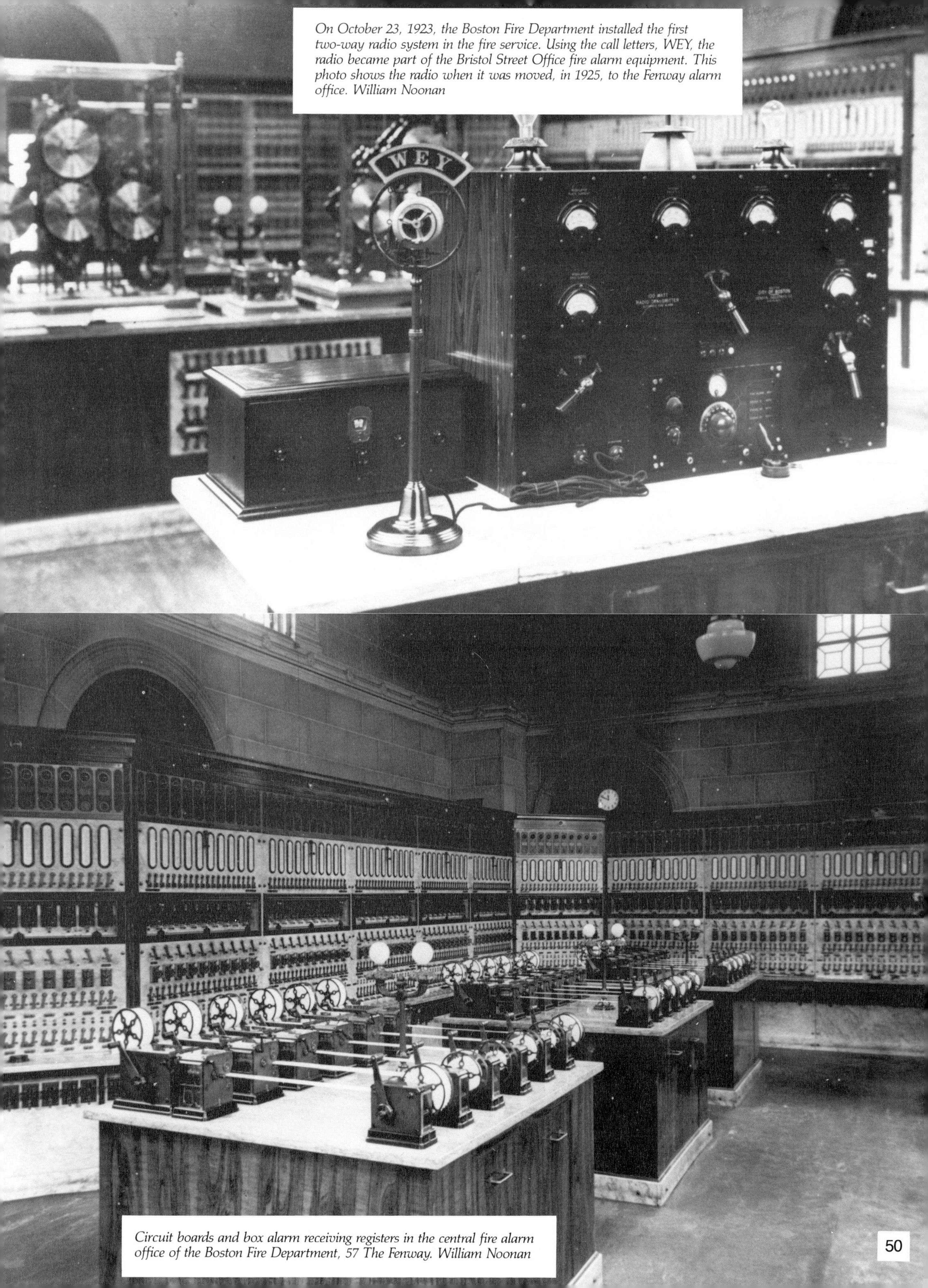

On October 23, 1923, the Boston Fire Department installed the first two-way radio system in the fire service. Using the call letters, WEY, the radio became part of the Bristol Street Office fire alarm equipment. This photo shows the radio when it was moved, in 1925, to the Fenway alarm office. William Noonan

Circuit boards and box alarm receiving registers in the central fire alarm office of the Boston Fire Department, 57 The Fenway. William Noonan

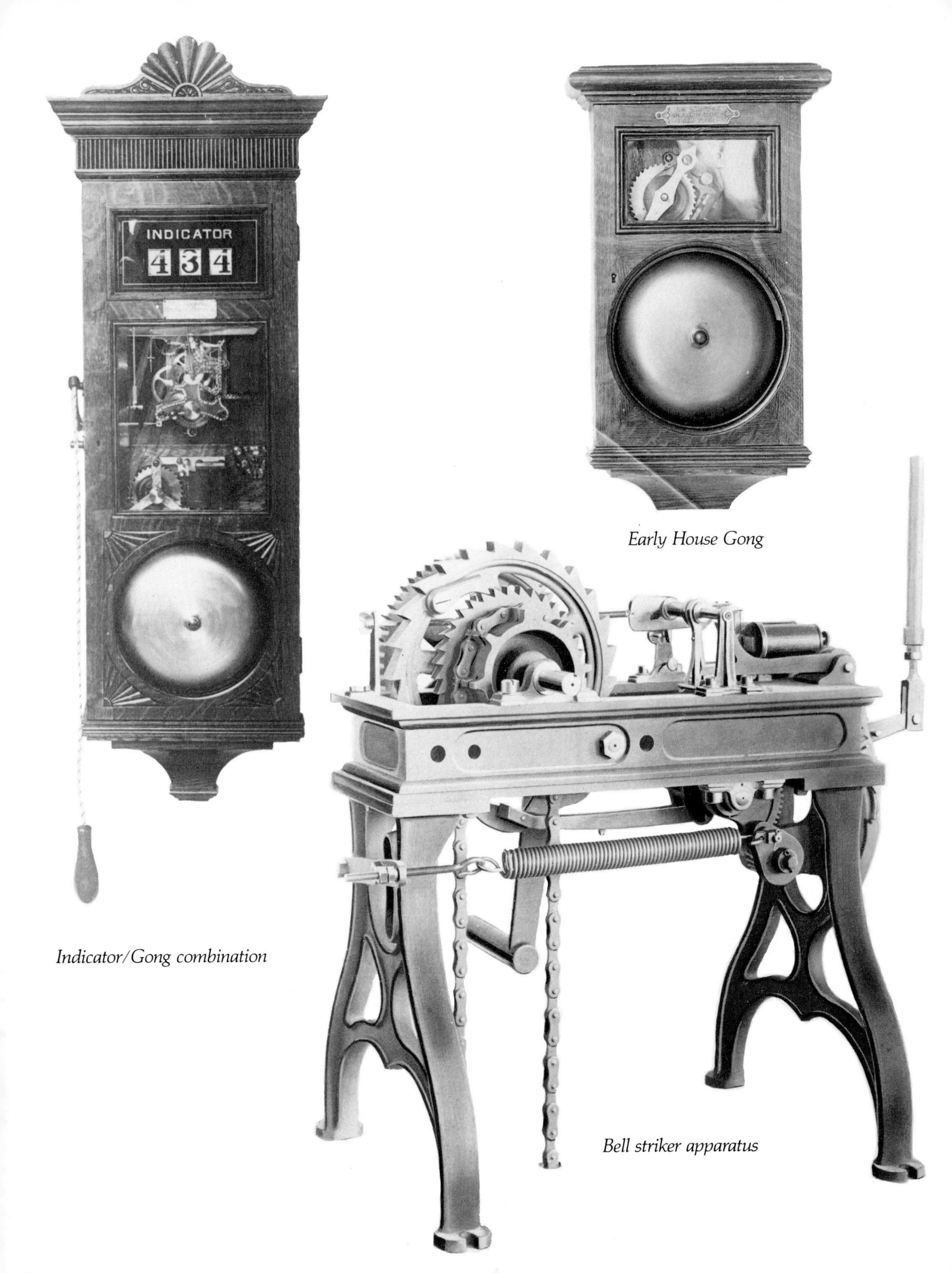

Indicator/Gong combination

Early House Gong

Bell striker apparatus

1965 — AUGUST 11. *Start of days of racial rioting in Los Angeles. More than 200 major fires were burning at the height of the rioting. Firebox alarms were ignored, because most were false or were pulled to lure firemen and police into dangerous areas and possible ambushes. Los Angeles was confronted with a growing false fire alarm problem before the Watts riots which only hastened the day when all fireboxes would be removed. Many cities followed suit: Not because of rioting, but because telephones far outnumbered alarm boxes and were more frequently used to report fires. Upkeep of the aging fire alarm system was another major factor. Among the holdouts are Boston, home of the first fire alarm system, and New York City.*

1967 — AUGUST. *United States Forest Services uses, for the first time under fireground conditions, an infrared camera mounted in a twin-engine airplane. The heat-seeking instrument penetrated smoke and clearly identified hot spots on strips of film. This use during the Sundance Fire in Idaho occurred while 26 million acres of wildland were burning in a 90-mile perimeter sprawling across five states. This method of fire detection and notifying ground-based firemen ultimately led to satellite identification of fires and transmission of alarms to the ground.*

1980 — 1990. *Profileration of hotel and multiple occupancy fire disasters in the United States and elsewhere results in laws requiring smoke detectors, heat detectors and better fire alarm alerting systems built-into these occupancies. Strong recommendations from fire service authorities for retrofitting old homes or including in the construction of new houses, built-in sprinkler systems which could reduce the national's annual fire deaths by as much as 50 percent.*

1982 — SEPTEMBER 4. *Early morning fire in the 55-year-old unsprinklered Dorothy Mae Hotel in Los Angeles kills 25 tenants. Casualties occurred despite the fact the building had smoke detectors, self-closing doors and other open stairway protection. Each room, moreover, had a telephone. The fire burned for at least half-an-hour before someone telephoned the fire department from an all-night convenience store. There were no fire alarm boxes in the city. The disaster stunned fire protection authorities and strongly suggested that sprinklers would have avoided the high loss of life. It also demonstrated that despite smoke and heat detection devices as well as telephones, some method must be developed for automatically turning in an alarm if occupants cannot, will not or are otherwise unable to do it themselves.*

1985 (circa). *Proliferation of 9-1-1 Emergency Telephone Reporting Systems. Enhanced systems automatically display the telephone number and location of the caller turning in the alarm. Hard and software in the system was a major breakthrough in calling for fire, police and emergency medical assistance. Enhancement did not eliminate the false alarm problem, especially among vandals and others who use pay telephones and vanish before the arrival of emergency services. That the 9-1-1 system's effectiveness was largely dependent upon alarm center personnel's capabilities was demonstrated by many lawsuits mostly traceable to untrained or incompetent alarm center personnel who failed to follow protocols, if any, with the result that there sometimes are delayed responses, notably to patients in severe medical distress, mostly due to heart attacks.*

1986 (circa). *Legally-required alarm systems in multiple occupancy hotels, apartment houses, condominiums, townhouses and private homes results in excellent built-in fire protection, but worse false alarm problems than during the heydey of street corner fireboxes. Fire authorities rightfully campaigned for these installations by pointing out that flagrant, malicious, accidental or mechanical malfunctions were preferable to one catastrophic loss of life. False alarm problems were compounded by the influx of many firms offering shoddy fire, police and emergency medical alarm systems that were triggered by homeowners and others unfamiliar with the system's operation, high winds and even household pets who set off purportedly highly-sophisticated systems. To firemen a call to respond to an Automatic Alarm invariably means an Automatic False Alarm.*

1989. *The Fire Department of New York, with more fire alarm boxes than any other city reports that it has 37,554 street fire alarm boxes. During that year the fire department answered only 5,842 fires from them, 2,259 other emergencies, and a phenomenal 29,453 false alarms. Balm for fire alarm box proponents in New York's report saying that 34,660 false alarms came from telephones.*

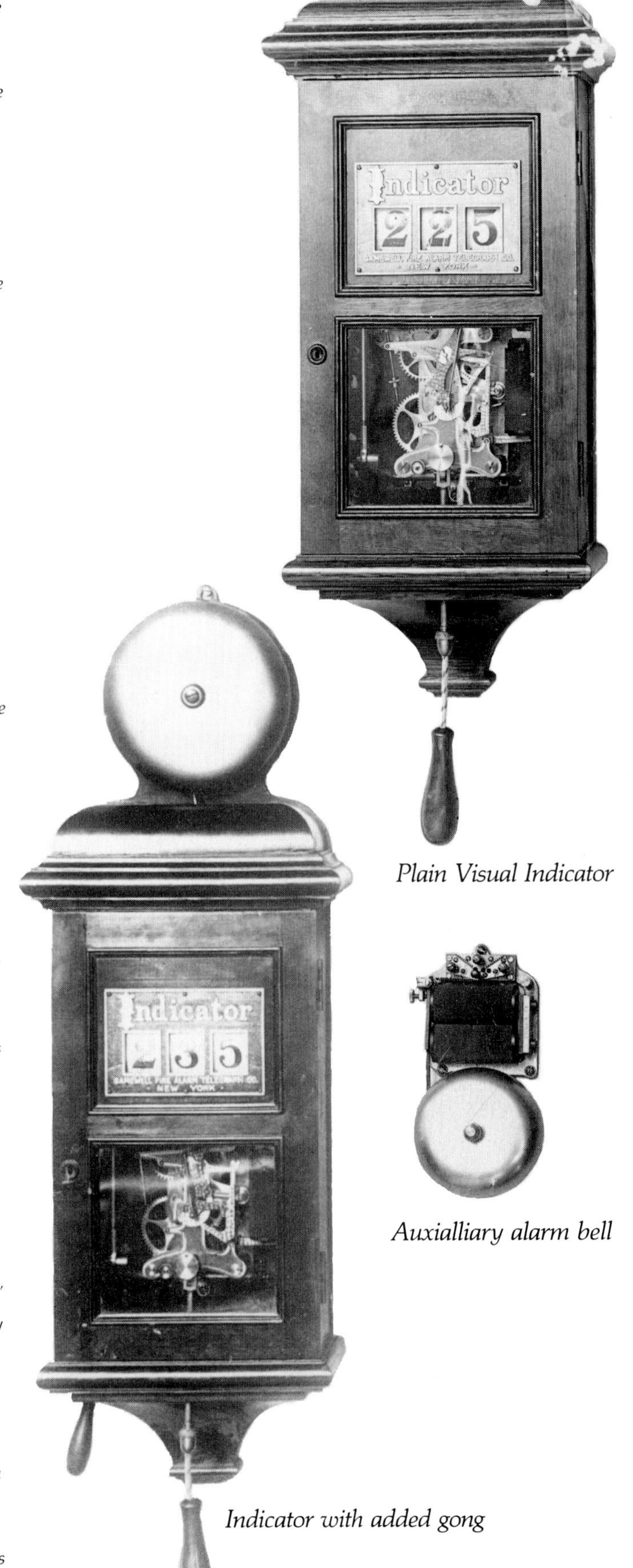

Plain Visual Indicator

Auxialliary alarm bell

Indicator with added gong

The Los Angeles Fire Department opened its Operations Control Division fire alarm office on June 26, 1973. The center, five floors underneath City Hall East, consolidated reception and transmission of alarms for the entire city and resulted in the closure of the Westlake Signal Office and satellite fire alarm offices elsewhere in the city. By 1990, the center was undergoing major renovations. Paul Ditzel

A COMPARISON OF SIZES

Early Gamewell Cottage Shell boxes varied considerably in size. From left to right: No. 4 Sector Box, No. 3 Sector Box, standard size box, oversized American Fire Alarm Telegraph Box.

FIRE ALARM
Telegraph Equipment

From the collection of (Ex) Chief W. Fred Conway

Left: Gamewell Keyless Door Box patented in 1875. Turning the handle did *not* send the alarm; it only rang a bell inside the box to alert passersby. The true alarm hook is inside the box, and many would-be reporters of fire failed to pull it, assuming the alarm was already sent when they heard the bell. Most cities removed the bells, but this box has the bell intact in good operating condition.

Many boxes, especially in larger cities, were mounted on pedestals with red lights to make them easy to spot at night. Smaller towns simply put the boxes on utility poles, on which they often painted a broad red stripe.

Top center: Gamewell 15″ House Gong. This was the most popular size, although Gamewell made smaller ones as well as a huge 18″ model. The beveled glass window is the result of Gamewell's predilection for exposing the mechanism as proof of its reliability.

These gongs are electromechanical, tripped by interruptions in the closed electrical alarm circuit, with powerful strokes which are spring-assisted. A large key keeps the spring wound.

Top right: Gamewell Plain Visual Indicator. As the alarm comes in, the number of the box appears in the glass window, giving a visual indication of the number in addition to the audible count of gong strokes. Indicators were often combined with the gongs into a single instrument. Gamewell alarm instruments were ornate, and were made of the finest materials by highly skilled craftsmen.

Above: Alarm Tapper. Likely from the Chief's Office or the watch desk of a fire station. ca 1860.

Gamewell Sector Box with Solid Door. 1880

Gamewell Sector Box with rare Half Glass Door. 1880

Gamewell Peerless Non-Interfering Sector Box with Cole Keyguard. 1916

Gamewell Peerless Successive Box with Quick Action Door. 1924

Gamewell Sector Box 1880 with optional Cole Keyguard added after 1906

Gamewell Excelsior Interfering Type Box with Smith Keyguard. 1900

Horni Fire Alarm Box. This Gamewell competitor later became "SAFA" Superior American Fire Alarm. ca 1930 — 1940

Harrington Three-Fold Box. Another Gamewell competitor ca 1960. This box, in mint condition, was never placed in service.

Top left: A rare Gamewell advertising specialty made of Gamewell's famous "Herculite" alloy. Is it a cigar/cigarette holder/ashtray? Chief Conway uses it on his desk for paper clips. ca 1930.
Top right: Brass key tag for a San Francisco fire alarm box ca 1870's. This key was kept by either a policeman, merchant, or prominent citizen who lived near the box, and it had to be obtained from its keeper before an alarm could be turned in. The obvious shortcomings of this procedure led to the invention of the Keyless Door in 1875.
Bottom left: One of the many Gamewell Fire Alarm Panels used in the Seattle, Washington alarm office from about 1913 to 1969.
Bottom right: Gamewell 8 Circuit Alarm Repeater used in Easton, Pennsylvania from 1893 to 1957. The oak cabinet contains weights which hang like a grandfather clock mechanism to supply the mechanical power to the electromechanical apparatus. Like the clock, it was wound with keys. After 1900 Gamewell made the cabinets out of steel, as some of the alarm offices feared the wooden cabinets constituted a fire hazard! The repeater automatically retransmitted fire alarms over all the alarm circuits in the city without the necessity of an attendant being on duty.

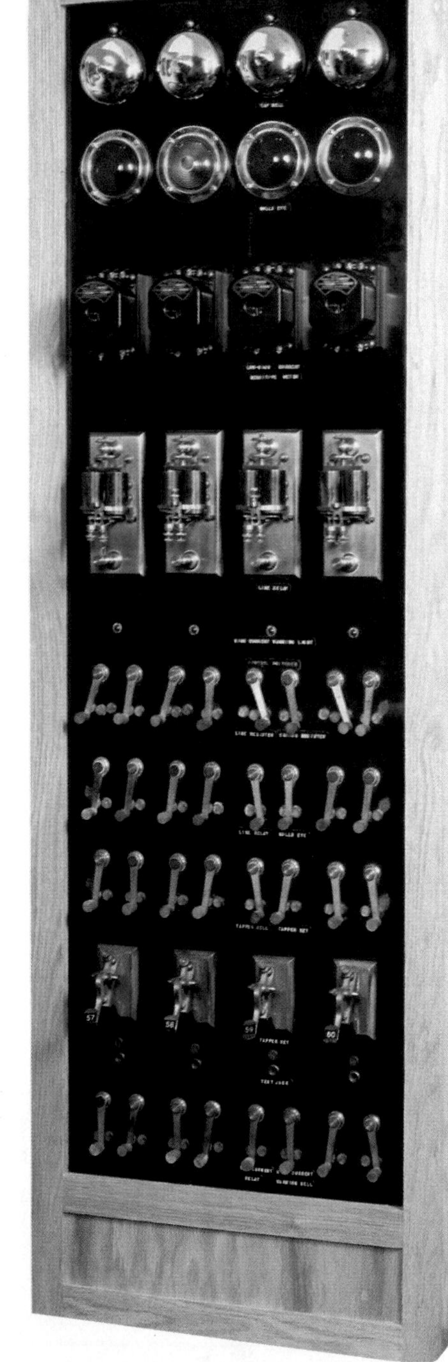

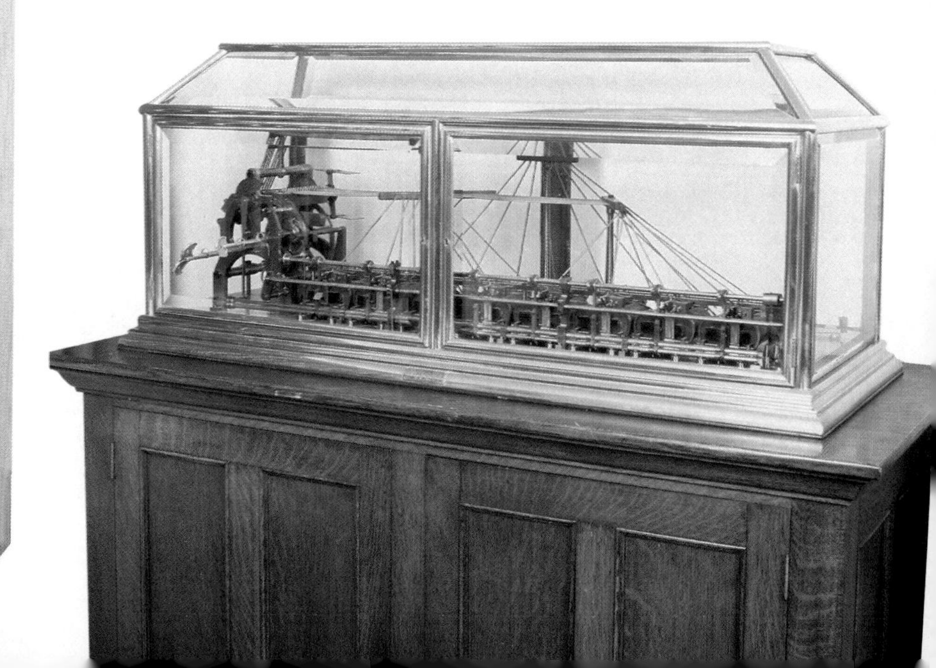

CHAPTER FIVE

CAVALCADE OF FIRE ALARM BOXES

From The Collection of John Bryan, Jr.
Photography by Dan Thacker

Early cast iron cottage style fire alarm box. This box is considerably larger than the more familiar fire alarm telegraph station clanging box with the typical fist grasping lightning bolts. Most early Gamewell boxes up to 1879 had the date of manufacture cast in the peak of the box. They were designated American Fire Alarm Telegraph, as manufactured by Gamewell & Company. These early boxes were manufactured in this size and a smaller size.

Early inner case for large American firebox. Early boxes and inner cases were painted black. Operating instructions were painted onto the door. Most early box movements were of the sector pull type: pulling the hook down would raise a weight that operated the clock-type mechanism or, in later models, wound a spring.

Interior view of a very early weight-driven fire alarm box with its very simple interbearing weight sector movement. Early boxes had steel covers protecting the clock-like movement. This box had the old style terminal block with a telegraph key and tap bell. Early boxes had the bell shunt switch located at the lower left corner to remove the tap bell from the circuit when the box door was closed. This was done to conserve battery power.

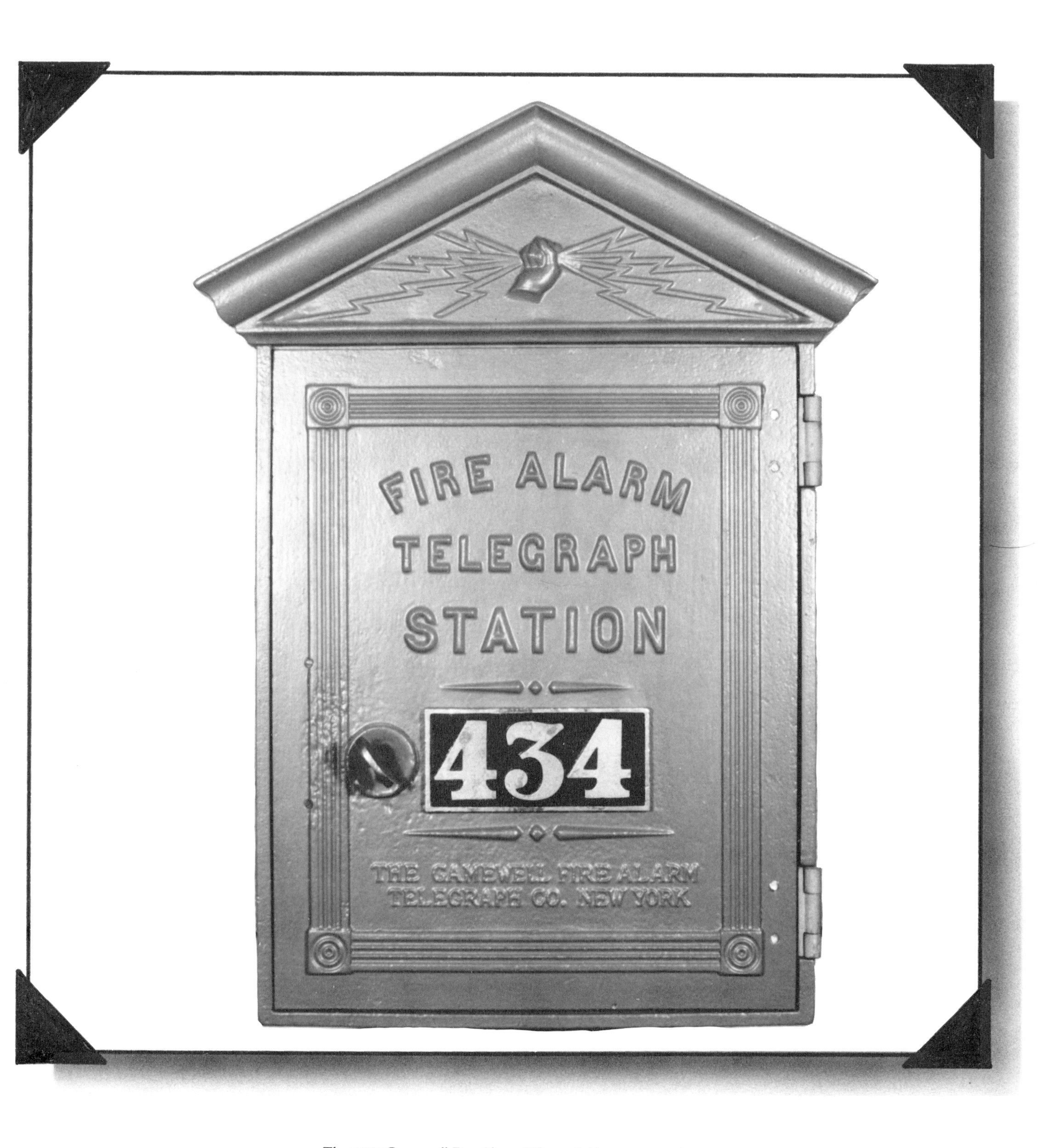

The 1879 Gamewell Fire Alarm Telegraph Company catalog illustrated a new fire alarm signal box design which, for the first time, adopted the familiar first grasping lightning bolts which became a registered trade mark of The Gamewell Fire Alarm Telegraph Company, New York. The new design no longer had the date of manufacture cast into the peak of the box and changed the wording from the earlier American Fire Alarm Telegraph Station to simply read "Fire Alarm Telegraph Station." This box was accessible only to those citizens and public servants who were assigned a numbered key which was used to open the locked door and sound the alarm.

Inner case of an 1880 Interfering Weight Sector Movement with its very simple weight-driven mechanism, old style hard rubber code wheel with key-break and a slate terminal block installed on a wood block for insulation from electrical overcurrent.

1880 style Fire Alarm Telegraph Station Box with a Smith Key Guard added. This key guard device was the first of its type where the break glass device was ***placed directly over the key slot*** *with the trapped key left in place in order to allow any person to transmit an alarm via the fire alarm signal box. During the period of these early boxes all door number plates were of the large cast brass type.*

In approximately 1880, the inner case instructions were no longer painted on the inner case door, but were now raised and cast directly into the door.

Very early non-interfering weight-sector box with its non-interfering coils located at the top of the movement bowl. This feature operated an electrical movement flow, which in the absence of current, such as when another box on the circuit was in operation, would prevent this box from sending its signal. This would require the user to activate the box a second or third time to send an alarm. However, most persons would not be immediately aware that the alarm was not being transmitted. Also, should this box be activated at the exact time as another box was closing the current while sending its signal, a jumbled signal would result, because both boxes would be operating on the same box circuit simultaneously.

Interfering Weight-Sector Movement with improved porcelain terminal block and bell shunt with a box test switch. This box was not equipped with a tap bell for communicating between the box and the fire alarm office.

Inner case of a later model Non-Interfering Weight Sector Box with its large old style code wheel resting on the key break, and a movement shunt just below the weight lever. This was provided to provide a higher level of protection from burnouts of signal contacts due to overcurrent from such sources as lightning, trolley and power lines.

Unusual Non-Interfering Weight Sector Box with an armature reset plunger located about weight level. This is very similiar to the famous Gardiner Non-Interfering box.

"Keyless Door," (Door Opening Type) of firebox. With the advent of providing easy access to the fire alarm box pull lever in order to reduce the delay in sending fire alarms, a new problem arose: false alarms. Charles Tooker of Chicago in 1875 patented a new door design known as the Keyless Door, which consisted of a large handle that would be turned to open the door, which, in turn, would ring a cast mechanical bell on the inside of the pot-bellied looking door. The doors were of two basic types: The Door Opening type, which required a citizen to open the door and then pull the mechanism pull lever. The second type was the Self-Starting type: Turning the handle to the right sounded the door alarm bell and also mechanically activated the inside mechanism pull lever, which caused the fire alarm to be transmitted without having to open the outside door. The small keyhole was to allow access to the box for service or testing without activating the door bell.

Inside look at Tooker's "Keyless Door" with its $4^1/_2$-inch mechanical bowl bell alarm. This bell was intended to attract attention to the fact that the fire alarm box was being pulled to send an alarm. A negative aspect was the fact that some people, upon hearing the alarm bell, would wrongly assume that they had successfully sent the fire alarm when, in fact, they had not. The self-starting door was designed to eliminate this objection. Some cities, such as Baltimore, removed these doors after a short period of time. Boston and other cities removed the bell but left the door intact.

In 1869, Moses G. Crane of Boston introduced a spring wound signal box with a non-interfering trigger pull lever which would not interfere with the operation of the signal box should an excited person pull the lever more than once to activate the box. This early box also contained a non-interfering feature which consisted of a pair of electrically operated coils at the top of the movement which would mechanically lock out contacts should power be absent on the circuit if another box was signaling with the circuit open between signal blows when this box was activated. This particular box is very early and was of the standard size, had a wooden terminal block, and was manufactured for Gamewell & Co. by Moses G. Crane of Boston.

Fire Alarm Telegraph Station Box with an unusual four digit cast brass door plate, Smith Key-Guard and a location light.

Latter version of a Moses G. Crane spring-powered non-interfering box. Note the large style code wheel with Brush-Break type signal contacts and brass box winding key (lower left).

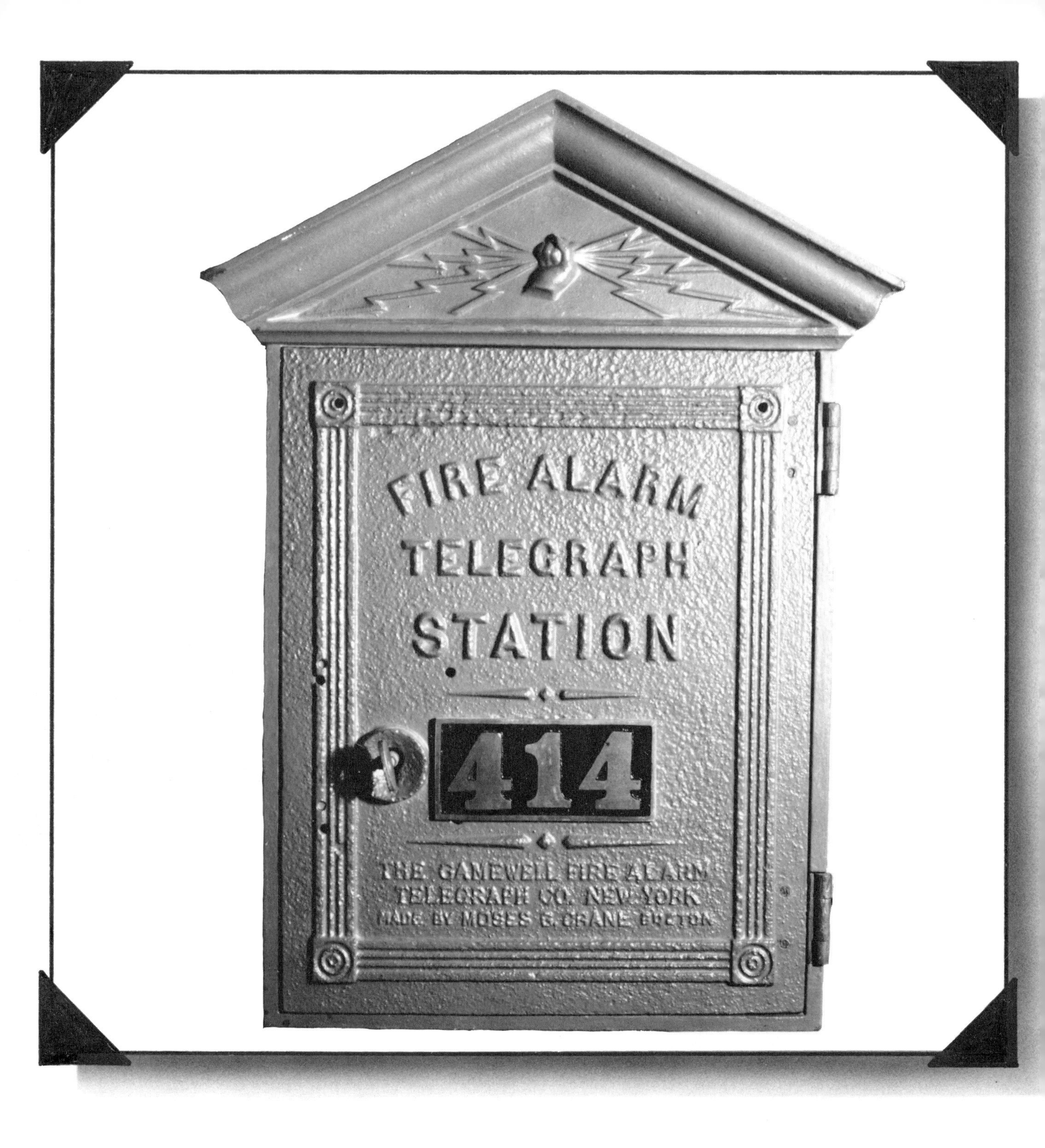

Early fire alarm telegraph station box, manufactured by Moses G. Crane, Boston, for the Gamewell Fire Alarm Telegraph Company.

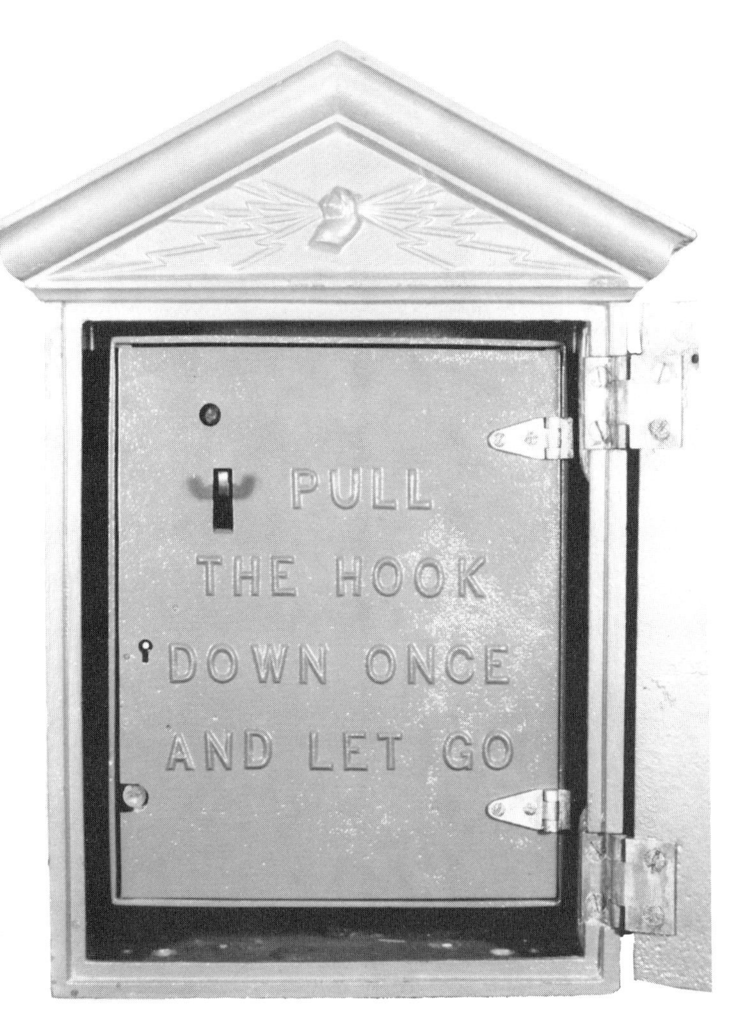

Fire Alarm Box Inner-Case, Gardiner Type, with non-interfering armature plunger hole at top left side of door above pull hook.

Inside of a 1916 Gardiner Type Non-Interfering Box. Few were ever in service.

Inside of a Gardiner Type Non-Interfering Box. Non-Interfering Coil with reset-armature is at the top of the movement bowl. When the outer shell door is opened, the reset pin is removed from the armature plunger, thus allowing the non-interfering coil to open and preventing this box from sending its signal should another box be pulled during the time it takes to open the exterior door and then pull the starting lever hook. This box was considered dangerous because it would not operate and send its signal when in the non-interfering mode; thus giving the citizen the impression that the box had in fact sent a fire alarm signal to the fire department. To reset the armature in order to send an alarm, the outer door would have to be closed to reset the armature and then the box operated once more. Should this procedure not be followed, the box would not send its signal regardless of how many times the excited citizen would pull the starting lever.

Early Instructional Door, circ 1900, with Smith Key Guard. In the 1890s more and more cities were giving the ordinary citizen access to the fire alarm box to sound alarms. Hence the need for the instructional door.

1916 brought many innovations and improvements to the fire alarm signal box, including the introduction of the familiar cast iron instructional box equipped with a Cole Key Guard.

Cast iron inner case door to one of the first Peerless Fire Alarm Boxes with a trigger pull handle to activate the box's main spring.

In 1916 Gamewell introduced the first Peerless Fire Alarm Box which was both positive non-interfering and successive. This box was not only non-interfering when pulled, but would also become non-interfering should an interfering box be operated by surrendering the circuit until the other box had completed sending its signal. The successive feature would allow the Peerless Box to test the circuit every revolution of the code wheel waiting until the circuit was again available to send its signal without interfering with another box on the circuit. This feature was a substantial improvement over the Ideal Non-Interfering Succession Box which would try for the circuit once every four rounds before shutting down after 8 to 16 rounds, depending on the model of the Ideal Box. This Peerless box was destined to become the forefather of today's three-fold box.

1916 Peerless Sector Box Inner Case with its Spring Sector Pull Hook.

1916 Peerless Non-Interfering Sector Box. This box would go four revolutions when the handle was fully operated and would wind the main spring. Should this box go into the non-interfering mode its signal would not be transmitted. If the box went into the non-interfering mode when operating, due to another box coming onto the circuit, the subsequent revolutions of the code wheel would not transmit signals. This required the citizen to realize after time had passed that the fire department was not responding and he would have to operate the box again. When in the non-interfering mode this box would operate mechanically but not electrically. The sound of the box operating would give the false impression that the signal had been successfully transmitted.

Instructional Door Box with Cole Key Guard in the open position to allow easy access to the tee-handle for opening the door. This outer case is diecast silicon-aluminum alloy; the trade name for which was Herculite. The Herculite box first appeared in use by Gamewell in 1928.

Outer shell of a Gamewell No. 3 Sector Box which was manufactured in the 1880s by Gamewell and is slightly smaller than the standard size cottage shell box. This box was in service in Baltimore County, Maryland.

Inner case of a Gamewell No. 3 Sector Box with its simplistic interfering sector pull mechanism, telegraph key and tap bell.

Outer shell of a Gamewell No. 4 Sector Box with an 1880 style solid door. This was a somewhat smaller box which was designed for smaller communities that did not need, or could not afford, the larger, more conventional style Gamewell boxes.

Outer shell of a Gamewell No. 4 Sector Box with the 1880 half-glass door design. Note the trap-lock and visibility of the Sector pull handle and operating instructions.

Outer shell of a Gamewell No. 4 Sector Box with a 1916 Instructional Door with the Cole Key Guard attached. The No. 4 Sector Box was manufactured from approximately 1880 to 1917-1918.

Inner case of a Gamewell No. 4 Sector Box with its sector pull handle, telegraph key and instructional movement bowl cover over a very simple interfering spring sector mechanism.

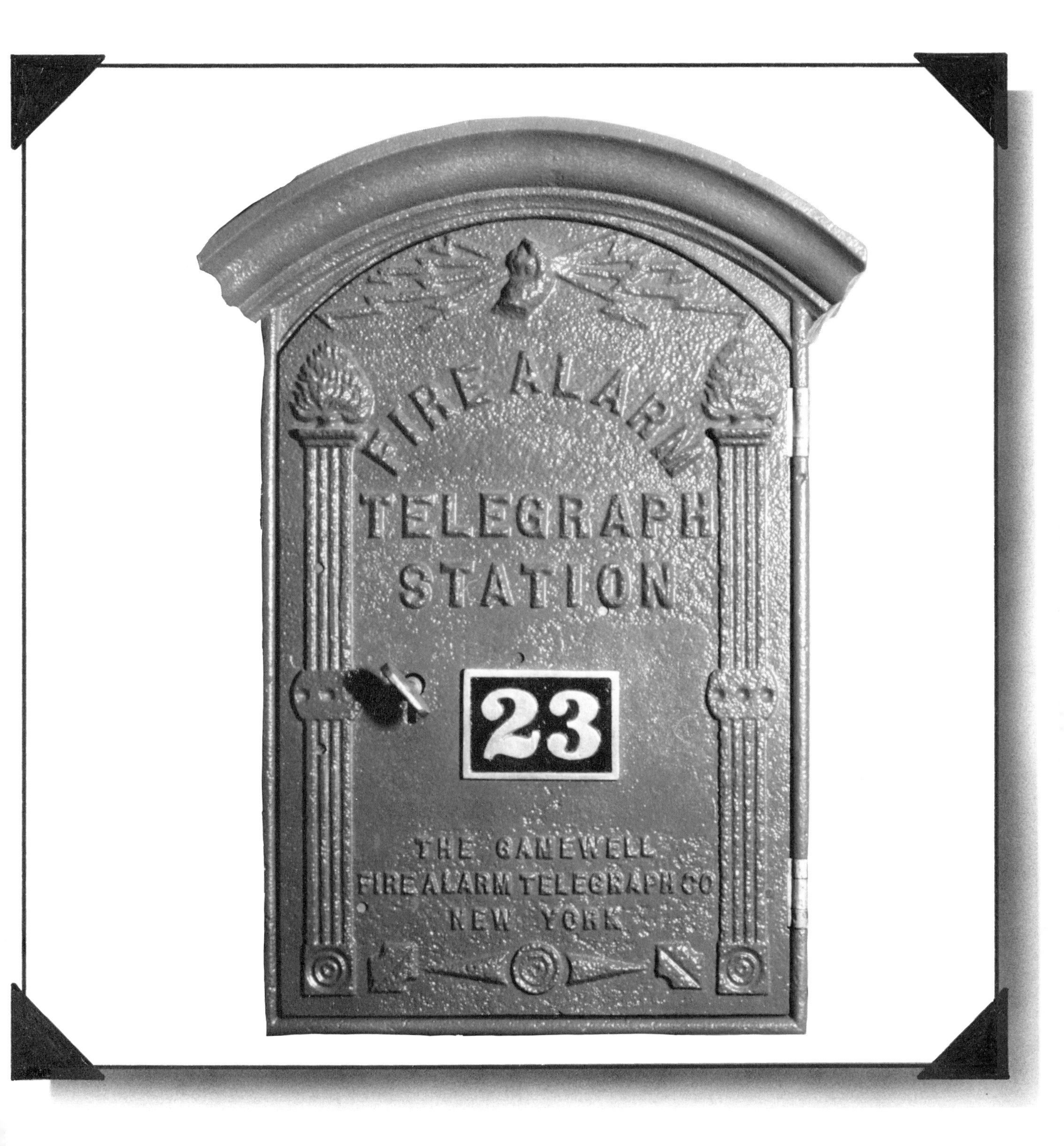

Outer shell of a Gamewell Excelsior Fire Alarm Box. Manufactured from early 1890s until 1917, these unique smaller boxes were offered for sale to smaller communities and were of interfering and non-interfering types. This style shell is featured in the Gamewell 1894 catalogue.

Gamewell Excelsor type fire alarm box equipped with a Cole Key Guard.

Later model Gamewell Excelsor Fire Alarm Box with a tee handle in lieu of a trap key lock for easy access to inner case starting mechanism. Excelsior box measures about 14″ × 10″ × 5″ and has a unique rounded style top with torch design columns similar to the earlier Utica Fire Alarm Telegraph Company boxes.

A Gamewell Instructional Door Excelsior Fire Alarm Box which was the last of its type to be manufactured in 1916.

Inner case of a Plain Interfering type Gamewell Excelsior Box. All Excelsior Boxes were of the Spring Sector Type.

Gamewell Excelsior inner case door with instructions for operating the Sector pull handle.

Gamewell Excelsior Non-Interfering Spring Sector Inner-Case. Note the Non-Interfering coil assembly in right corner of the mechanism bowl assembly which operated on the Gardiner non-interfering principle. These inner cases were very similar to those of the Utica Fire Alarm Telegraph Company.

Outer shell of an unusual Pierce & Jones Fire Alarm Telegraph Box, New York. During the early years of fire alarm companies flourished only to go out of business or were purchased by Gamewell.

Interfering Spring Sector Inner Case of a Pierce & Jones Fire Alarm Telegraph box.

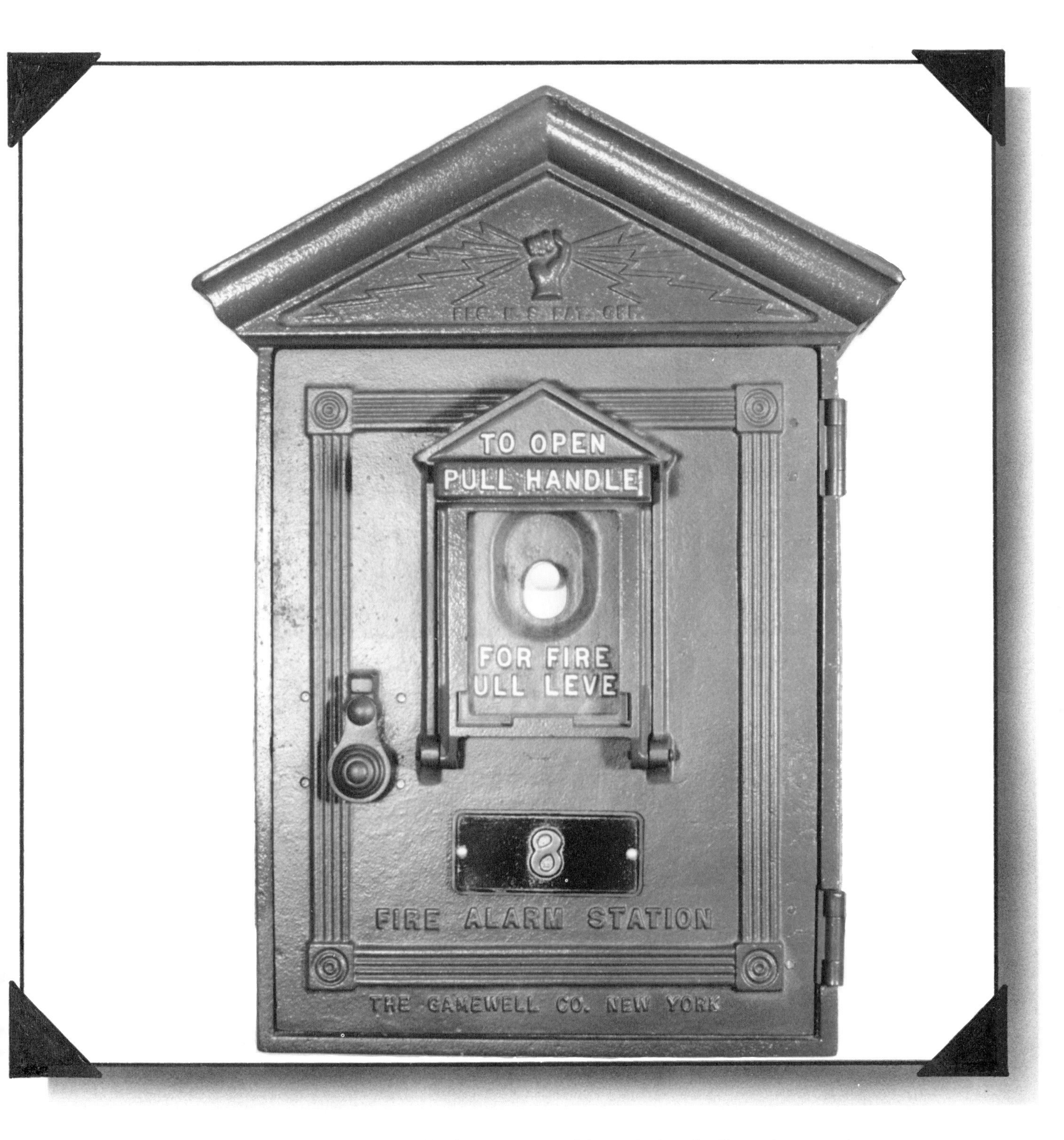

Gamewell's first "Quick Acting Door" was introduced in 1922. George Alfred Jackson made application for a patent on December 27, 1922. His patent was granted and numbered 1,479,608 on January 1, 1924. This new door design gave the citizen, for the first time, easy access to the fire alarm box starting lever. The door consists of a smaller door with a cast iron, bottom hinged, frame containing a panel of clear glass. This openable door protected the starting lever from the weather and visually exposed the starting lever and its instructions. The purpose of the clear glass panel was to eliminate the problems associated with Key-Guard boxes by having to first break the glass, open the outside door and pull the starting hook down once. The clear glass panel usually exposed the starting lever for all passerbys to see, thus eliminating problems associated with break-glass type doors, such as those concealing the starting lever and users being injured by broken glass. All a citizen now had to do was grasp the door handle, pull it down, and pull the starting lever. The clear glass design was to last only until 1924. This door design also incorporated a new lock design manufactured for Gamewell by Yale Lock Company. The lock was changed to a newer type design which was opened by a flat key carried by fire department personnel only, and is still in use today. The new Quick Action door design at its introduction, did not eliminate the familiar key guard equipped boxes, but was soon destined to do so.

In 1924 Gamewell revised its all glass cast iron framed quick action door with a new style instructional solid cast-iron door provided with a convex Bullseye site glass which visually exposed the fire alarm box starting lever.

Inner case of the First Peerless (Positive Non-Interfering Successive) Gamewell Fire Alarm Box which was introduced in 1916. This style inner case can be found in both the Instructional Door Cole Key Guard equipped boxes and in earlier versions of the Quick Action Door box.

In 1922 Gamewell introduced its improved Peerless Fire Alarm Box with its red porcelain enameled steel inner case which replaced the old cast iron inner case.

1922 Improved Peerless (Positive Non-Interfering Successive) fire alarm inner case with its distinctive white porcelain enameled steel mechanism bowl housing the movement.

In 1928 Gamewell revised its Peerless fire alarm inner case design. The inner case was changed to Bakelite: a brownish red plastic that is unique to this 1928 design only. This new design relocated the lightning arrester to the outside of the inner case, and fused it against overcurrent. This fuse arrangement is again unique to the 1928 design. Note the lack of an inner case lock because the public was no longer given access to the inner case due to the new Quick Action door design of the outer shell.

Inside view of a 1928 design Gamewell Peerless box with its new design and its dust proof glass bowl covering the mechanism back board, new tap bell arrangement, telegraph key and test block assembly.

1928 brought about a change in the outer shell of the Gamewell fire alarm signaling box. The construction of the shell was changed from the traditional cast iron to the much lighter aluminum alloy shell with the trade name of Herculite. This new style box weighed about 25 pounds which was approximately one-third the weight of the older cast iron boxes. In addition, the quick action door was provided with a spring-loaded hinge to shut the door after opening.

In 1931 Gamewell introduced its "Three-Fold" fire alarm box which utilized the famous Automatic Grounding circuit with its ability to go to ground to complete the circuit should the line loop be interrupted due to a break. The inner case was now white Herculite aluminum. The fused porcelain lightning arrester was eliminated in favor of a non-fused Bakelite arrester. The Bakelite arrester/terminal bar is unique to the 1931 Three-Fold design. This new box was provided with a distinctive diagonal white stripe on the sides of the outer shell to visually identify the box.

Inside view of Gamewell 1931 Three-Fold inner-case with its brass movement works, which was very similar to the 1928 Peerless except for the addition of the automatic grounding feature located at the top of the mechanism. Three-Fold meant that the box was non-interfering, successive, and had the automatic grounding feature. This new three-fold principle remains in use into the 1990s.

In 1939 Gamewell made some slight changes to the Three-Fold fire alarm box. The tap bell was changed to an upright position. The lightning arrester/terminal bar was changed and the mechanism was no longer brass but polished steel. The use of fire decals began to be used on the outer shell to visually identify the box.

Gamewell 1951 style Three-Fold fire alarm box. During that year Gamewell changed their design to a modern styling with a slightly curved peak outer shell and rectangular plastic sight glass in the Quick Action door. This new shell design was no longer a single cast shell, but was now of a three-piece design to allow easy mounting of the box on a pole. There were no notable changes in the inner case except that over time more aluminum and plastic mechanism parts were utilized.

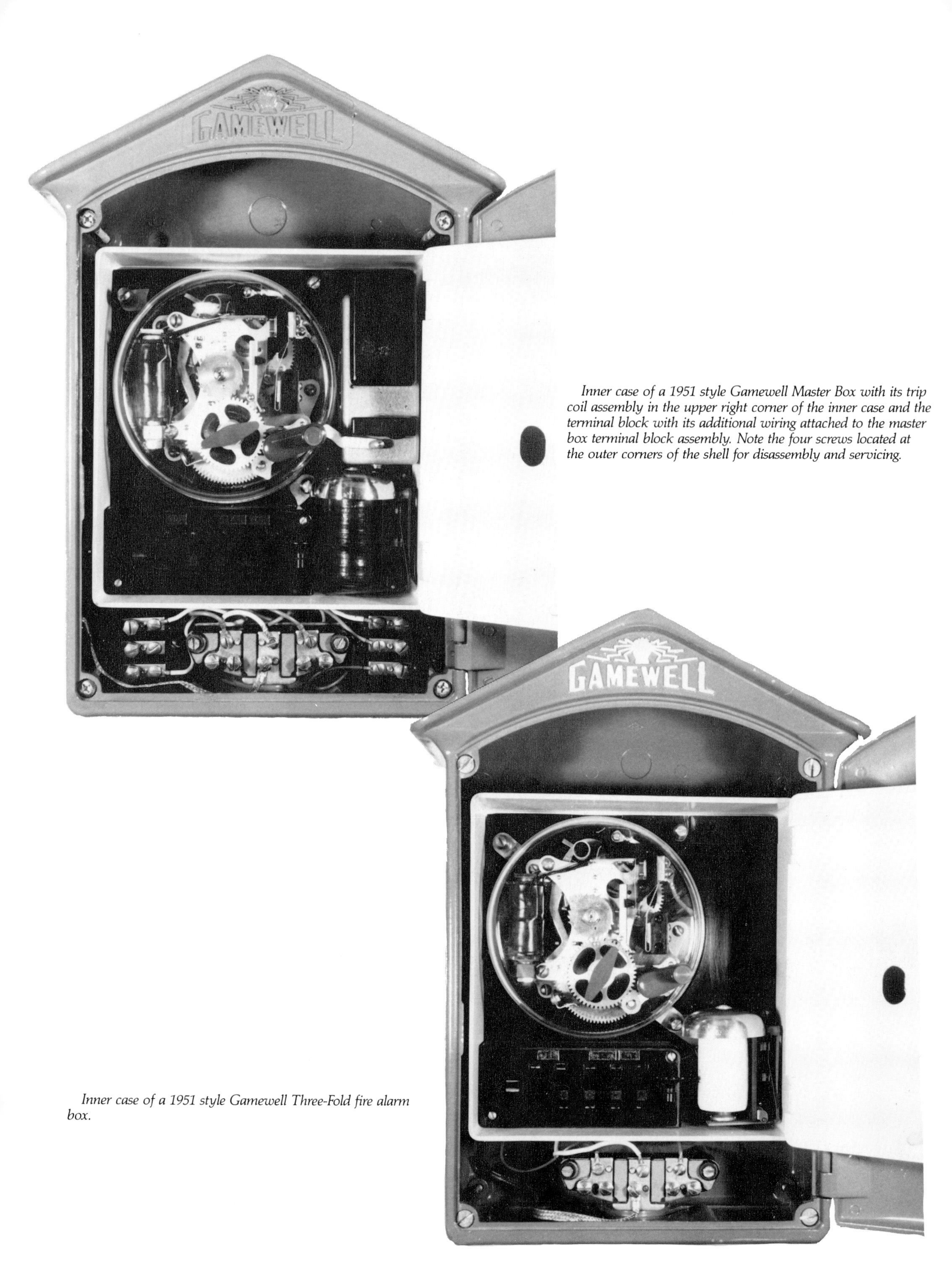

Inner case of a 1951 style Gamewell Master Box with its trip coil assembly in the upper right corner of the inner case and the terminal block with its additional wiring attached to the master box terminal block assembly. Note the four screws located at the outer corners of the shell for disassembly and servicing.

Inner case of a 1951 style Gamewell Three-Fold fire alarm box.

Unique 1951 style Gamewell combination Three-Fold fire alarm box with an emergency telephone inside for calling police, ambulance or other public safety services.

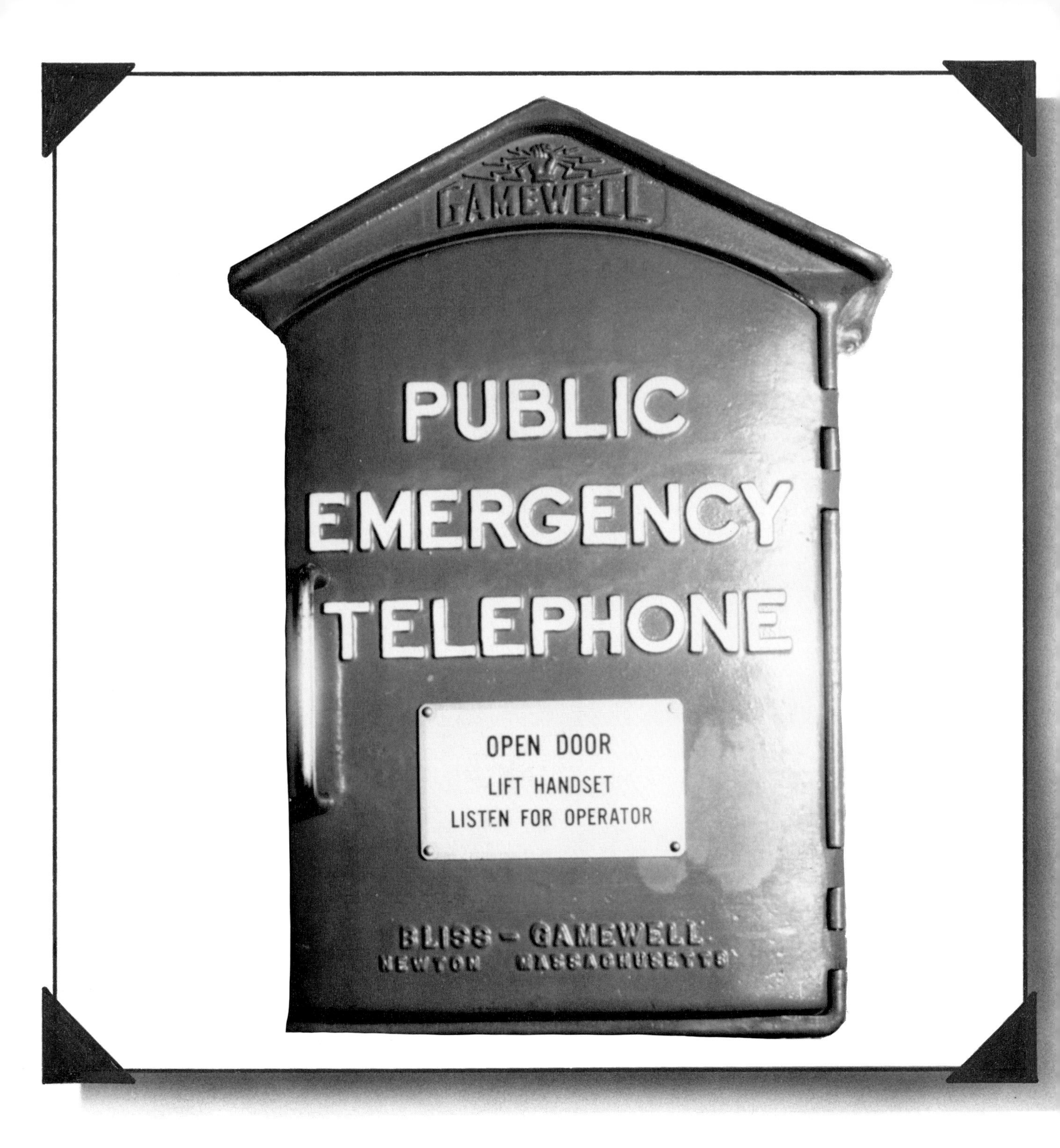

Gamewell 1951 style Public Emergency Telephone for reporting a fire, calling police or an ambulance.

APPENDIX A

GONGS, PEDESTALS, AND VARIOUS FIRE ALARM TELEGRAPH INSTRUMENTS AND ACCESSORIES

Gongs from Gamewell catalogs from 1895 to 1910. Left is an ornate combination gong/indicator. As the number of the box banged out on the gong, the box number showed in the glass window above — a visual as well as audible indication of the box number.

Both the clockwork mechanism and the walnut or oak cases were made of the best materials with high quality workmanship. Gamewell displayed the mechanism inside glass windows on the gongs and indicators as proof of their dependability.

The gongs are electromechanical, meaning that each tap on the gong by the hammer is a strong blow because of a spring that is kept wound with a large key. These gongs will awaken the soundest-sleeping firemen on the second floor of the fire house, and send them sliding down the pole within seconds. But the sudden clanging of the gong in the middle of the night precipitated many heart attacks. Today soft tones or chimes are used.

Bottom left is a small 6″ Excelsior model gong. On the right are 8″ and 15″ firehouse gongs. The brass gongs were kept highly polished, and were gleaming showpieces lending a mystique to the firehouses of yore.

BELL STRIKING MACHINE

An Acme bell striking machine as advertised in an old Gamewell catalog. Prior to the advent of the Gamewell compressed air-driven Diaphone horn, bell strikers were in common use. In fact, a 1980 Gamewell press release states, "Believe it or not, this striker still had widespread use in the early 1940's, and some may still be operational."

Although the first bell striking machine is attributed to Farmer, bell strikers are associated with Moses Crane, a Gamewell associate, who perfected them and was issued many patents on them.

One model of Crane's bell strikers could strike an incredible 10,000 blows with a single winding, the weight dropping only 21 feet.

THE GAMEWELL COMPANY

AUTOMATIC REPEATERS

When a system becomes large enough to be divided into two or more circuits, automatic repeaters are used for the purpose of repeating the signals from any circuit over all the others.

When a system is so divided, the necessity for non-interference devices for preventing interference and confusion of signals through boxes being operated on two or more circuits at the same time becomes readily apparent.

Our repeaters are so designed that all circuits are operative from any of the box circuits, and are equipped with positive non-interfering devices which absolutely prevent any interference between circuits. The repeaters are so arranged that, should a break occur in any circuit, the instrument will, after causing one blow to be struck on all the alarm apparatus in the system, automatically lock out the disabled circuit, leaving the rest of the system intact, and when repairs have been made to the disabled circuit the repeater will automatically take it into service.

Repeaters are provided with box and alarm circuits, the idea being to have the alarm apparatus installed on separate and distinct circuits so that an accident to any box circuit will not interfere with or disable any part of the alarm equipment.

Repeaters are provided with indicating devices for the purpose of indicating the circuit from which an alarm is coming in. The repeating contacts are of the highest grade of material, flexibly mounted with suitable tension to insure positive operation of the circuits. Our repeaters are constructed throughout of the highest grade of material and skilled workmanship which can be obtained for the purpose.

TEN CIRCUIT AUTOMATIC SWITCHBOARD WITH REPEATER SWITCHES

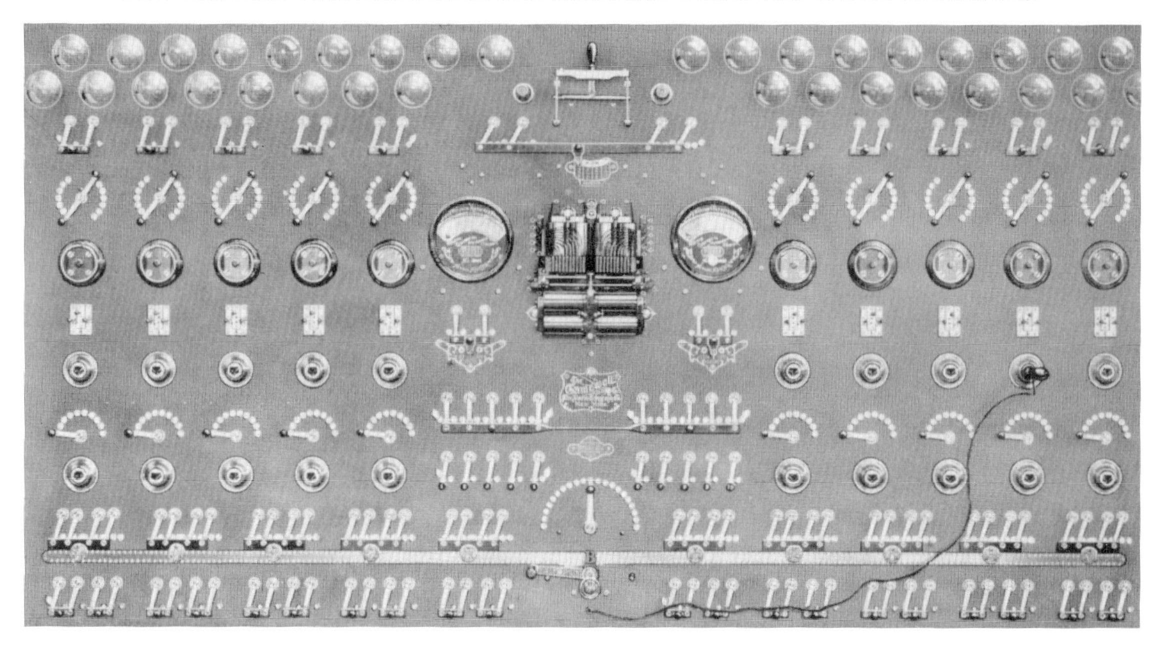

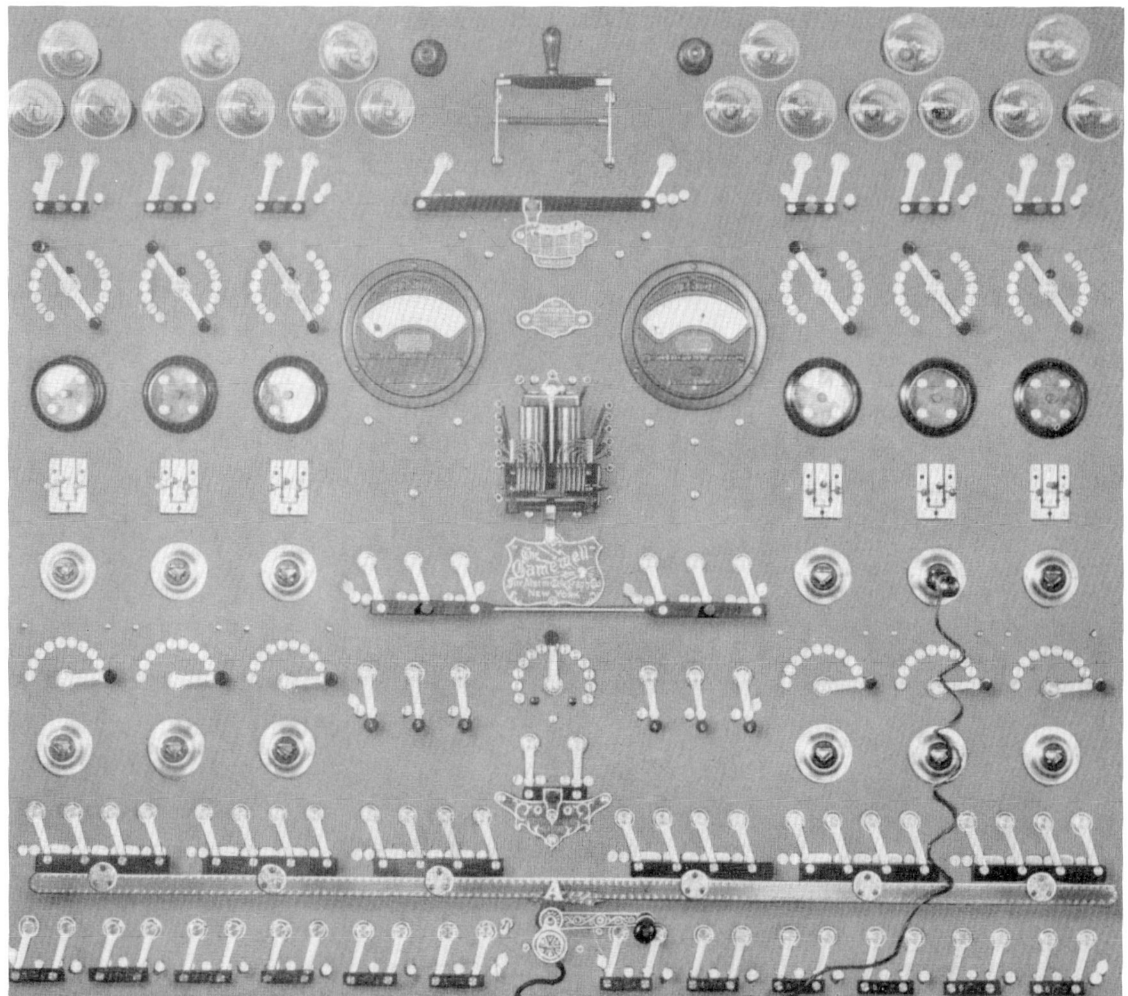

SIX CIRCUIT MANUAL SWITCHBOARD WITH REPEATER SWITCHES

Nonpareil Puncturing Registers

Ideal Punch Register

Excelsior Time and Date Stamp

Peerless Take-up Reel

A new reel with distinct improvements

GAMEWELL TURTLE GONG

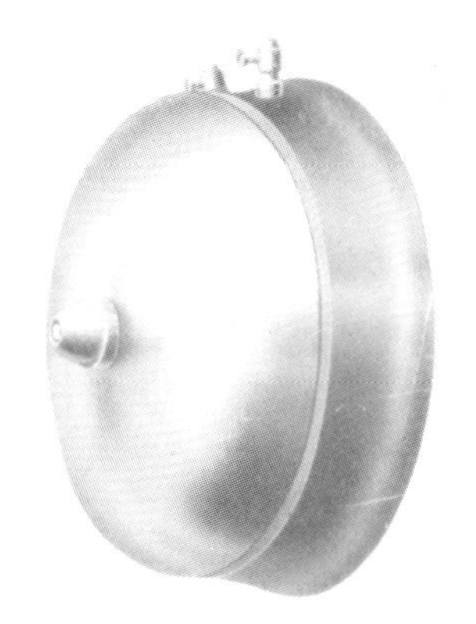

6″ TURTLE GONG

Uniform Time Relay

UPRIGHT CALL BELL

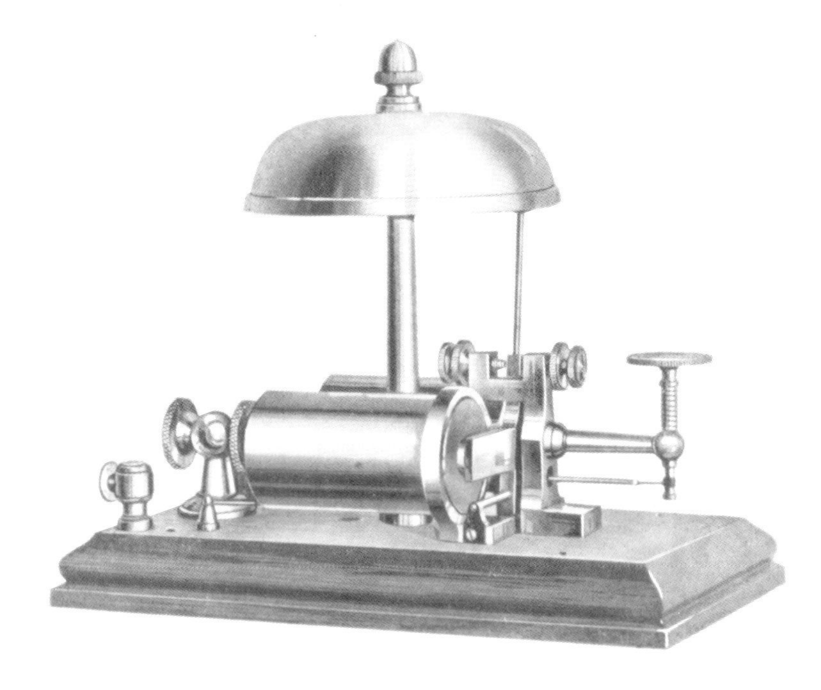

6″ TORONTO TAPPER

BOX LINE RECORDING SET

Consisting of Punch Register, Time Stamp with Clock, Peerless Take-up Reel, and Shelf with Brackets and Line Terminals

PEERLESS TWO-SPEED TRANSFORMER

FIRE ALARM CENTRAL STATION EQUIPMENT
PEERLESS TRANSMITTER

TRANSMITTING AND MASTER RECORDING OUTFIT—CLASS A SYSTEM WITH TWO PEERLESS TRANSMITTERS

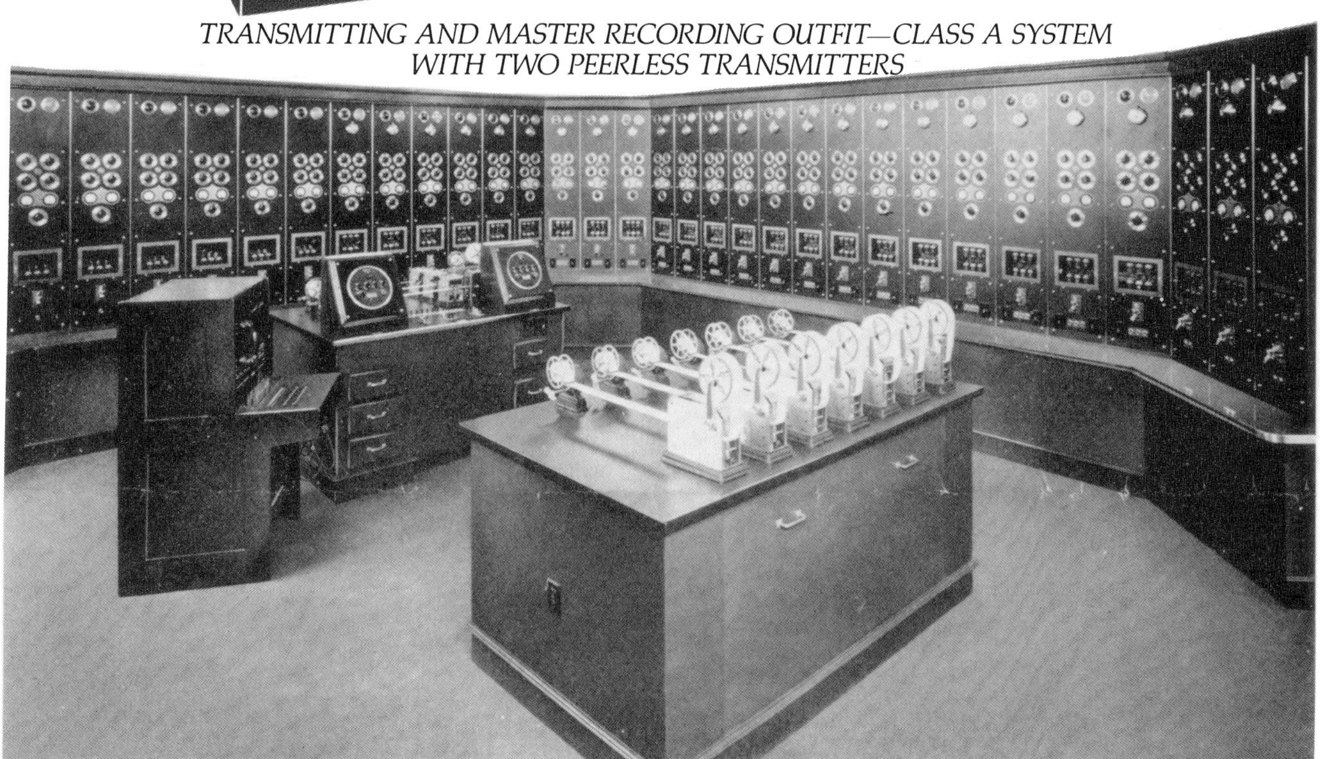

FIRE ALARM CENTRAL STATION. THREE-FOLD TYPE

SACRAMENTO, CALIFORNIA

GAMEWELL DIAPHONE

(Compressed Air Firefighters' Alerting Horn)

A FIRE ALARM THAT REALLY SAYS FIRE

AUTOMATIC WHISTLE BLOWING MACHINE

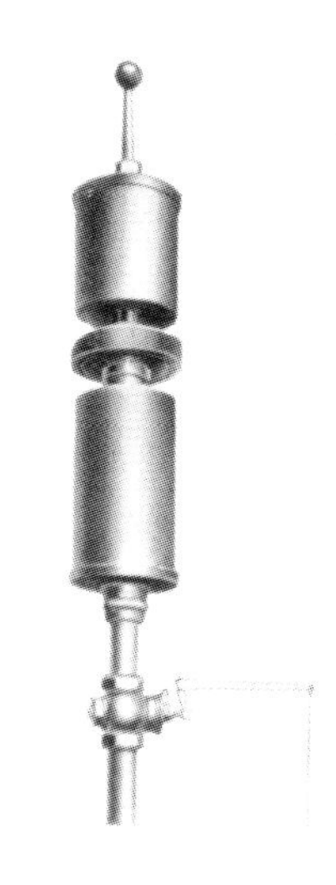

Steam Gong and Balance Valve shown in upper part of illustration not included unless specified

UMBRIA CLOCK

Umbria Clock with Electrical Contacts for Testing Fire Alarm System

Automatic Light Switch

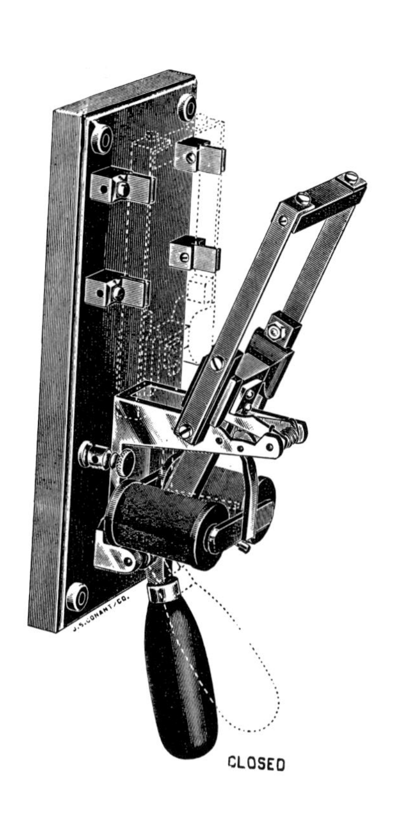

GAMEWELL PEDESTALS

FOR

Fire and Police Signaling Systems

Gamewell Pedestals are made in four classes to meet the varied needs of different cities.

TYPE "A" PEDESTALS

These posts are cast iron throughout, sturdily built and most attractive. Terminal space for two to thirteen circuits is provided in the base. When regular terminal box is used, a porcelain terminal block is provided in the box for four additional circuits. The ground extension is six inches in diameter and goes about two feet into the ground. Ground extension nine inches in diameter, or even larger, can be supplied if desired, to care for large cables. Two openings for laterals four and one half inches wide and seventeen and one half inches high are provided in the ground extension. Additional openings can be provided when larger pipe is used. This type of post may also be furnished without ground pipe, but with a flange inside the base for bolting to sidewalk.

TYPE "B" PEDESTALS

The flange, bowl, box holder, and acorn are cast iron. The top post and ground extension is one piece of wrought pipe. The same sized ground extension for lateral openings and terminal facilities is provided as in the Type "A" Pedestals.

TYPE "C" PEDESTALS

These posts are cast iron throughout and much simpler in design than the preceding types of posts. A porcelain terminal block for eight circuits is mounted in the terminal box, and no terminal facilities are provided in the base. The posts, which have a cable box in the back, Nos. 720 to 725, inclusive, will care for twenty-four circuits, but provisions can be made for any number of circuits up to ninety-four. These posts are supplied in two styles, the plain top post and the split type top post. (The top post extends from the sidewalk up to the box holder.) In case a split type top post is broken it is not necessary to disconnect wires from terminal block in order to install new post. The cable and terminal block can be left intact while change is being made.

TYPE "D" PEDESTALS

These posts are similar in general construction to the Type "B" Pedestals, but slightly different in artistic design. The same sized ground pipe openings for laterals and terminal facilities are provided as in Types "A" and "B."

SPECIAL

Certain older types of pedestals are not now standard but can be furnished on special orders.

General Information

Globes — The globe holders are fitted with pipe and socket to take any standard lamp. Globes are six inches in diameter at base and either ten or twelve inches at largest diameter. Red and green are the standard colors, although any color can be obtained.

Prices on these Pedestals do not include fire alarm or police boxes, but do include terminal boxes with hinged door and lock when this type of post is ordered. Terminal blocks, globes, and lamps are not included in prices. All the pedestals are painted with a coat of rust-preventive paint only.

When ordering Pedestals please give catalogue number and specify the type of fire or police box to be used, i.e., Peerless Box, Ideal Box, Seven Call Police Box, etc.

The Gamewell Company

NEWTON UPPER FALLS, MASS.

Type "A" Pedestals

Front View

No. 601

Side View

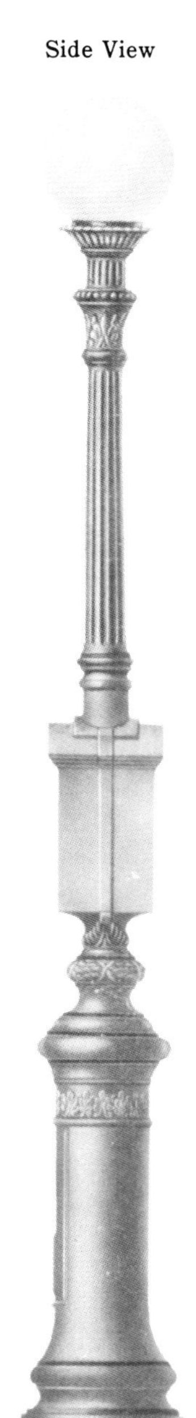

No. 601

View showing Police Box

No. 628

Side View

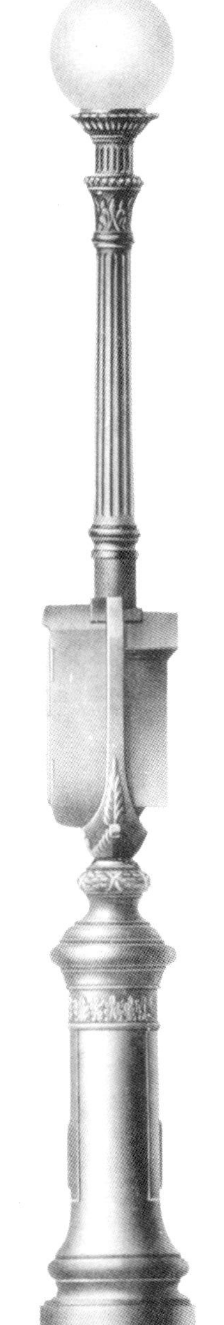

No. 628

View showing Fire Alarm Box

No. 628

Fire Alarm Box and Terminal Box Pedestal

No. 601 Peerless, Ideal, or Gardiner box with terminal box.

No. 601C Same as above, with two lights.

No. 601D Same as above, with three lights.

No. 616 Standard police box, with terminal box and one light.

No. 628 Peerless, Ideal, or Gardiner box with standard police box, with one light.

No. 629 Same as above, with two lights.

No. 630 Same as above, with three lights.
Can be furnished with either red or green globes.

Can be adapted for Excelsior or Sector Fire Alarm Boxes or Exemplar Police Box

Type "A" Pedestals

Front View Side View

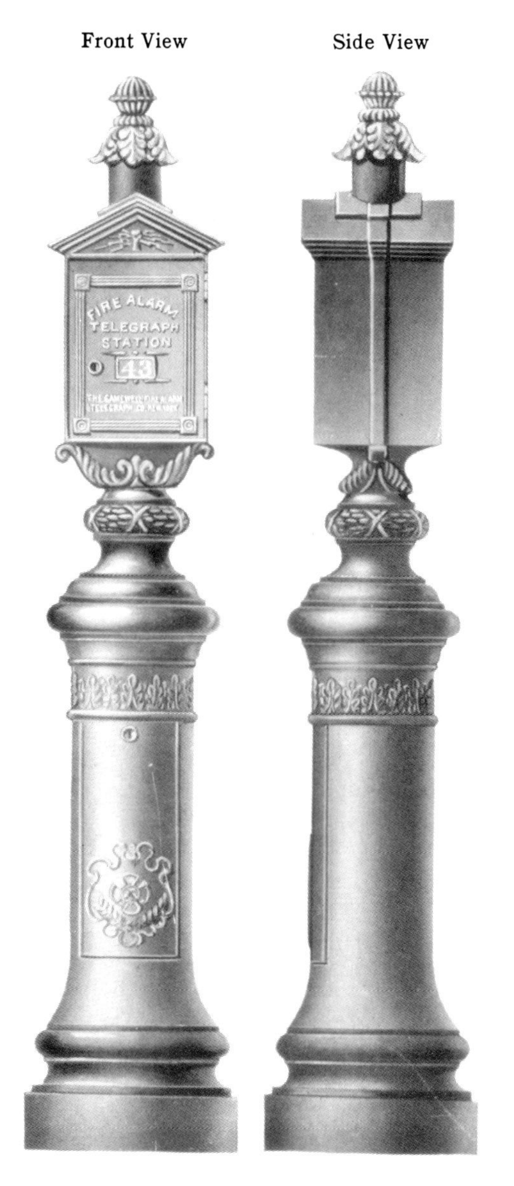

No. 625 No. 625

Fire Alarm Box and Terminal Pedestal

No. 625 Peerless, Ideal, or Gardiner box with terminal box.

No. 626 Standard police box with terminal box.

No. 627 Peerless, Ideal, or Gardiner box with Standard police box.

Type "D" Pedestals

Front View

No. 602

Side View

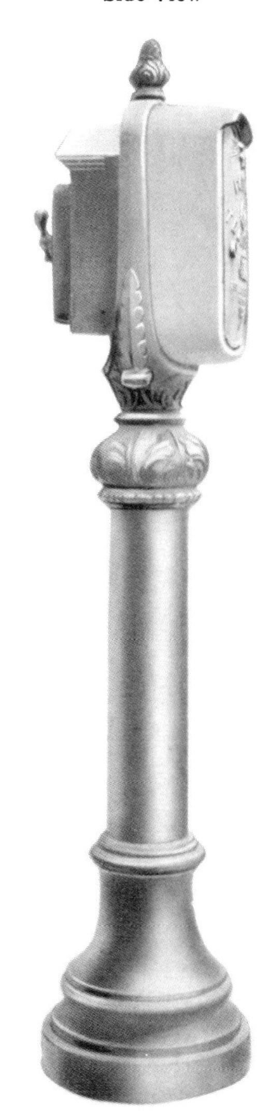

No. 603

Fire Alarm Box Pedestal

No. 602 Peerless, Ideal, or Gardiner box. No cable box. Terminal space in base.

No. 911 Standard police box. No cable box. Terminal space in base.

Fire Alarm and Police Box Pedestal

No. 603 Peerless, Ideal, or Gardiner box with Standard police box. No cable box. Terminal space in base.

Bolts and nuts for fastening box to pedestal furnished with Nos. 602, 911 and 603.

Can be adapted for Excelsior or Sector Fire Alarm Boxes or Exemplar Police Box

Type "B" Pedestals

Front View

No. 617

Side View

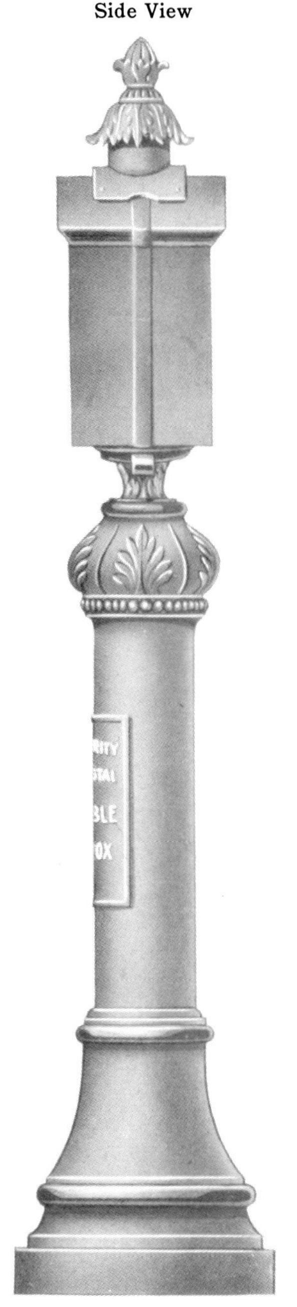

No. 617

Alarm Box and Terminal Box Pedestal

No. 617 Peerless, Ideal, or Gardiner box with terminal box.

No. 618 Standard police box with terminal box.

No. 636

Terminal Post

No. 636 Terminal post for test purposes.

No. 637 Same style post as No. 636,except that base is same as shown on No. 625, page 3, Type "A."

Can be adapted for Excelsior or Sector Fire Alarm Boxes or Exemplar Police Box

Type "B" Pedestals

Fire Alarm Box and Police Box Pedestal

No. 619 Peerless, Ideal, or Gardiner box with Standard police box

Can be adapted for Excelsior or Sector Fire Alarm Boxes or Exemplar Police Box

Type "B" Pedestals

Front View	Side View	View showing Police Box	Side View	View showing Fire Alarm Box

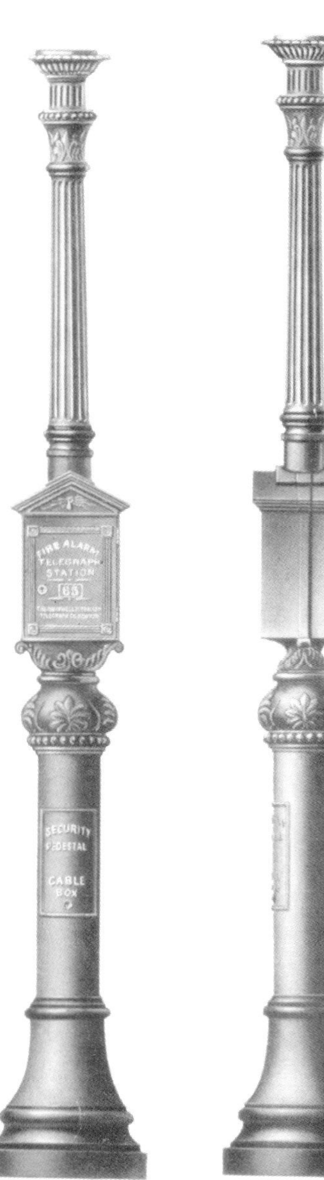

No. 620 No. 620 No. 622 No. 622 No. 622

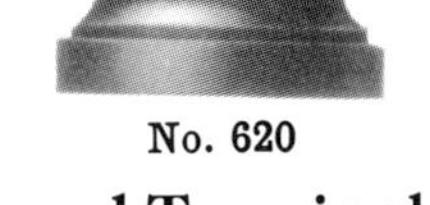

Fire Alarm Box and Terminal Box Pedestal

No. 620 Peerless, Ideal, or Gardiner box with terminal box.

No. 621 Standard police box with terminal box.

Fire Alarm Box and Police Box Pedestal

No. 622 Peerless, Ideal, or Gardiner box with Standard police box.

Can be adapted for Excelsior or Sector Fire Alarm Boxes or Exemplar Police Box

Type "B" Pedestals

No. 623

No. 624

Fire Alarm Box and Police Box Pedestal

No. 623 Peerless, Ideal, or Gardiner box with Standard police box, with column for two lights. Space for terminals provided in base.

No. 624 Peerless, Ideal, or Gardiner box with Standard police box, with column for three lights. Space for terminals provided in base.

Can be adapted for Excelsior or Sector Fire Alarm Boxes or Exemplar Police Box

Type "C" Pedestals

No. 700

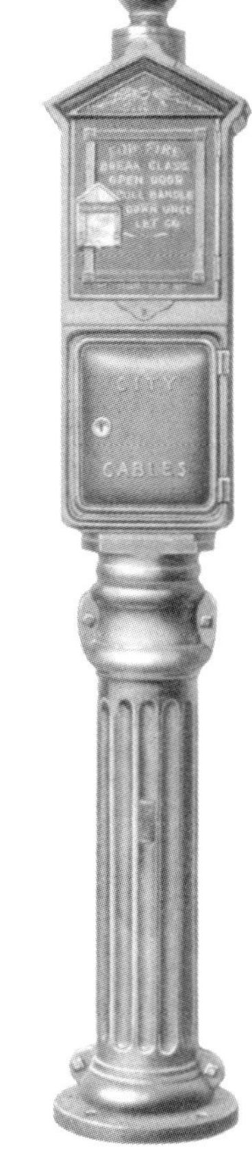

No. 703

Plain Top Post

No. 700 Peerless, Ideal, or Gardiner fire alarm box, with cable box below.

No. 706 Same as No. 700, except that it has shelf for fastening in place of leaded joint.

No. 720 Same as No. 700, with long cable box with door on back.

Split Type Top Post

No. 703 Peerless, Ideal, or Gardiner fire alarm box, with cable box below.

No. 723 Same as No. 703, with long cable box on back.

No. 708

Plain Top Post Shelf Support

No. 708 Peerless, Ideal, or Gardiner fire alarm box, with cable box below.

No. 707 Same as No. 708, except ladder rest and mushroom top in place of globe.

Type "C" Pedestals

No. 702

Plain Type Top Post

(Illustration at left)

No. 702 Peerless, Ideal, or Gardiner fire alarm box, with cable box below.

No. 701 Same as No. 702, except ladder holder and mushroom top in place of globe.

No. 722 Same as No. 702, with long cable box with door on back.

No. 721 Same as No. 702, except ladder rest and mushroom top in place of globe, and with long cable box with door on back.

Split Type Top Post

(Illustration at right)

No. 705 Peerless, Ideal, or Gardiner fire alarm box, with cable box beneath.

No. 704 Same as No. 705, except ladder rest and mushroom top in place of globe.

No. 725 Same as No. 705, with long cable box with door on back.

No. 724 Same as No. 705, except ladder rest and mushroom top in place of globe, and with long cable box with door on back.

No. 705

Type "C" Pedestals

No. 730

(*Illustration at left*)

No. 730 Standard police box.

No. 732 Same as No. 730, with column and globe as shown on page 9.

No. 731 Same as No. 730, with column, ladder rest, and mushroom top.

Plain Top Post

(*Illustration at right*)

No. 740 Combination fire and police box. Standard police box, with either Peerless, Ideal, or Excelsior fire alarm box.

No. 742 Same as No. 740, but with column and globe as shown on page 9.

No. 741 Same as No. 740, but with column and ladder rest.

No. 740

Type "C" Pedestals

No. 750 Cable Box, sixty-circuit capacity.

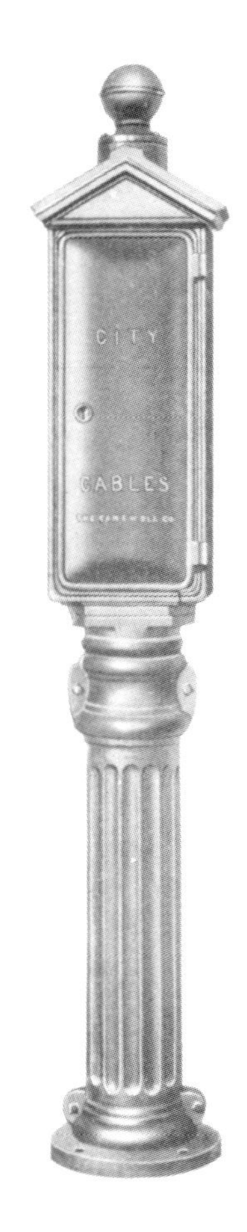

No. 753 Cable Box, with split type top post, sixty-circuit capacity.

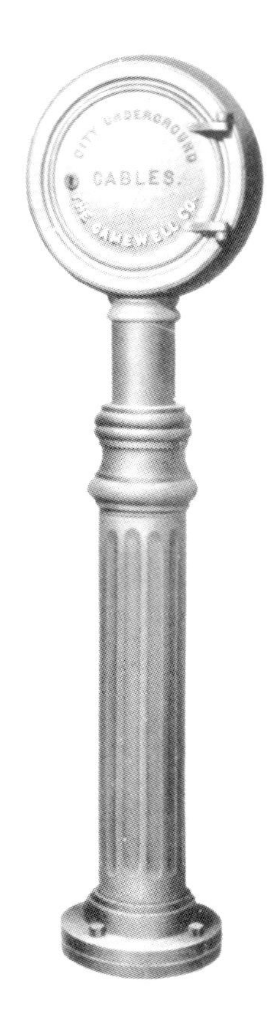

No. 604 Banjo type cable box, thirteen-circuit capacity.

Various Parts of Pedestals

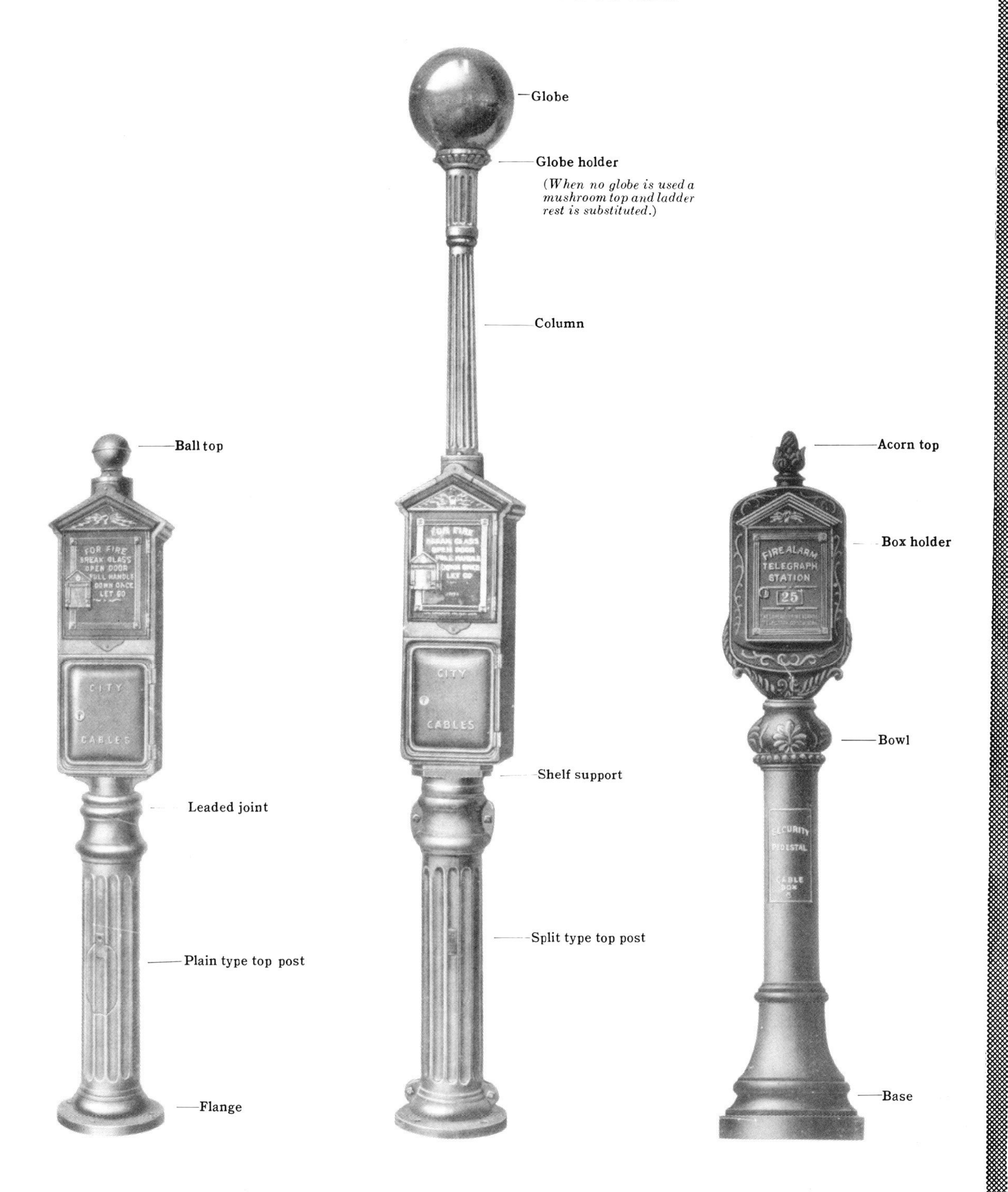

Ground extension is the section from the sidewalk down

APPENDIX B

RUNNING CARDS

Every fire station and alarm office kept running cards which specifically showed which engines, hook and ladders and other equipment was to respond on first or greater alarms. The cards frequently showed those companies which would cover areas left unprotected when a second or greater alarm was sounded. If companies due on the first alarm were not available, dispatchers usually picked the first company or companies shown on the second alarm assignment.

546

ENGINE COMPANIE

10 13 37

48 29 7 6

32 24 25 27

42 35 16 17

3 4 34 11

21 41 15 22
5 33 23 36
2 18 9 8
26 12 39 38
20 49 30 46
47

2141 **Aliso & Los Angeles Sts.**

	ENGINE				TRUCK		SAL	SQD	H.U.	CHIEF	
										B.C.	A.C.
1st	**4**	**3**	**24**	**9**	**4**	**3**		**4–9**		**7**	
2nd	**9**	**6**	**28**		**24**	**28**			**3**	**1**	**1**
3rd	**30**	**8**	**70**								
4th	**18**	**45**	**7**								

COMPANIES CHANGE LOCATIONS

	ENGINE					TRUCK	
2nd	**52–3**	**50–4**	**31–9**			**29–28**	**17–24**
3rd							

F-682—55M—10-60

Running card for downtown Los Angeles firebox. George Bass

1245 So. Van Ness (Bet. 23rd & 24th)

Nearest St. Box **5511** 23rd & S. Van Ness

VICINITY BOXES

5455 5469 5472
5473 5477 5511
5512 5525 5526 5534

SE DER	CHIEFS A	CHIEFS B	TRUCKS	TANK W.	S.C.	R.S.	F.B.	CHIEFS AND COMPANIES TO CHANGE LOCATIONS: ENGINE COMPANIES				CHIEFS	TRUCKS
	A3	6	7-11		4							SPECIAL COVER-IN ONLY	
		2-11	9		1			21-13			41-37		4-7
		10	6-8					15-29			22-27		13-8
7		5	1					1-25			5-35		
		3	3					33-42	23-17	36-34	44-43		10-

RESERVES	A	B	TRUCKS	TANK W.	S.C.	R.S.	F.B.
	A2	4	4 13		2	1	1
		9	10 5		3	2	
		1	12 17				
	A1	8	15 2				
		7	19 16				
			18 14				

SPECIAL APPARATUS

RESPONSE	SL	AC	SS	FW	VW	CW	FMW	LT	ST
SPEC. CALL	1	1		1	1	1	1	1	3
2ND ALARM			1						
RESERVE	2							2	4
SPEC. ORDERS					(6)				

San Francisco Fire Department running card. Willy Dunn

6316 Wilson & Milwaukee Aves. 4600-N. 5200-W.

ENGINES							TRUCKS		SQUADS		H.P.	W.T.	Ambulance	Div. Mar.	BATT. CHIEF	
			7	69	94	108	23	58		11				6	22	30
			86	91	124	125		38		6	7		7			20
			43	56	89	117		56		4						24
		76	96	106	114	128										
		12	42	68	79	111										

COMPANIES TO CHANGE LOCATION

Alarms	ENGINES					TRUCKS		BATT. CHIEF	
Second	67-94	71-108	77-125	79-124	128- 69	13-58	43-23	3-20	23-22
Third	6-43	12- 91	21-117	33- 89	42- 56		36-38		8-24
Fourth		30- 69	44-114	98-106	111- 7				
Fifth			23- 56	105-124	109- 68				

SEPTEMBER, 1958

Running card for upper northwest side of Chicago Fire Box 6316. Paul Ditzel

691 Park Ave. and 86th St.

ENGINE COMPANIES							H & L CO'S		DEP CHF	BATTALION CHIEF		B T	W T	TO SUPPLY FUEL	From Depot No.	COMPANIE… ENGINE COMP…		
22	44	89					13	~~39~~ 16	5	10	11							
39	53	56	74				~~16~~ 26			8				W3 ~~10~~ 9	~~16~~ 11			
76	58	35	40	8		8	~~26~~ 43			12				W2 ~~1~~	16	16-22 [44]	~~80-58~~ 26-74	6…
23	36	65	47	21		21	~~25~~ 25 ~~43~~			9				W 5	16	~~26~~ [18]-23	50-36	

1751-09 (B) 300 July 1, 1909

Independence Hall, Philadelphia, home of the Liberty Bell, had a running card with an appropriately-numbered Box 1776. Paul Ditzel

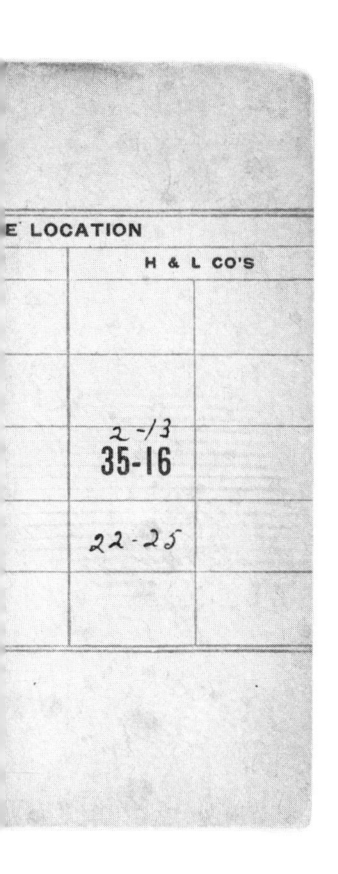
E LOCATION
H & L CO'S
2-13
35-16
22-25

Fire Department of New York running card for Box 691. Changes of response districts required firemen to make the necessary corrections directly on the card. This card originally was issued on July 1, 1909. Steven Scher

1776- S.E. Cor. 6th & Chestnut Sts.

81

INDEPENDENCE HALL

SECOND FLOOR - CONTROL ROOM WEST WING

ENGINE COMPANIES	LADDER COS.	BATT. CHIEFS	RESCUE COS.	BOAT COS.
8, 20, 11, 4	2,23	4,5		
17,(21) 3, 1,(43) 13	5	6	1	
	(9)			
10, 29,(53)(24) 2, 27	1	3		
49, 44,(34)(6) 31, 47	11			
60, 5, 25, 50				
45, 28, 16, 65				

( ) = BY PASS COMPANY

H.P. 1ST ALARM

PHILADELPHIA FIRE DEPARTMENT

76-73 FIRE BOX LOCATOR

APPENDIX C

Fire Alarm Equipment Manufacturers

There were around three dozen manufacturers of fire alarm boxes and related telegraphic equipment. The following is a representative compilation of those companies as provided by Chief Robert N. Noll, Yukon, Oklahoma, Fire Department.

Akron Electrical Manufacturing Company, Akron, Ohio

American Fire Alarm Telegraph Company, Boston, Massachusetts

L.W. Bills Company (Louis W. Bills), Lexington, Massachusetts

Charles T. and J.N. Chester Company, New York

City of Chicago Fire Department, Chicago, Illinois (A local foundry made their shells and they used a commercially available movement.)

City of San Francisco, Department of Electricity, San Francisco, California (They used their own outer shell or Universal Tool and Manufacturing made a box for them.)

City of St. Louis, Missouri, Fire and Police Telegraph Department, St. Louis, Missouri (One of many cities that manufactured their own fire alarm telegraph boxes and used a commercially available movement.)

Cregier Signal Company, Chicago, Illinois

Faraday Electric, New York, New York

Federal Signal Company, Blue Island, Illinois

Foote, Pierson, & Company, New York, New York

Fyr-Fighter Company, Orrville, Ohio

Gamewell Fire Alarm Telegraph Company, New York, New York

The Garl Signal Company, Akron, Ohio

Gaynor Electric Company, Louisville, Kentucky

General Signal Company, Fletcher, Ohio

Harrington-Seaberg Company, Moline, Illinois

Harrington Signal Company, Moline, Illinois (Harrington-Seaberg was this company's parent.)

Holtzer-Cabot Electric Company, Boston, Massachusetts

Horni Signal Manufacturing Corporation, New York, New York

Inter-State Fire Alarm Company, Omaha, Nebraska (Manufactured the Richmond Fire Alarm System.)

Loper Fire Alarm Company, Stonington, Connecticut

McFell Signal Company, Chicago, Illinois

Moses G. Crane & Company, Boston, Massachusetts

Municipal Fire and Police Telegraph Company, Boston, Massachusetts

New York City Fire Department (A city that manufactured its own shells and used a commercially available movement.)

Northern Electric Company (This company was a Gamewell licensee and manufactured fire alarm boxes in Canada.)

Pierce & Jones Fire Alarm Telegraph Company, New York

Southern Switch and Signal Company, Shreveport, Louisiana

Star Electric Company, Binghamton, New York

Sterling Siren Fire Alarm Company, Incorporated. (It is thought that this manufacturer was also known as or did business as The Interstate Machine Products Company.) Rochester, New York

George M. Stevens Company, Incorporated, Boston, Massachusetts

Superior American Fire Alarm & Signal Company, Meriden, Connecticut (Formerly General Signal Company.)

Union Fire Alarm Company, New York City, New York

United States Fire and Police Telegraph Company, Boston, Massachusetts

Universal Tool and Manufacturing Company, San Francisco, California

Utica Fire Alarm Telegraph Company, Utica, New York

Western Electric Company, Chicago, Illinois, and New York City

Fire AND Police Telegraph Systems

Home Office, BOSTON, MASS.

1897

The following are 60 of the Cities and Towns using apparatus of the United States Fire & Police Telegraph Company design:—

Aurora, Ill.
Astoria, Oregon.
Bath, Me.
Boston, Mass.
Battle Creek, Mich.
Barre, Vt.
Cambridge, Mass.
Columbus, O.
Cleveland, O.
Crompton, R. I.
Dallas, Tex.
Davenport, Ia.
Danvers, Mass.
Eau Claire, Wis.
Fitchburg, Mass.
Grand Rapids, Mich.
Hartford, Conn.
Janesville, Wis.
Jersey City, N. J.
La Crosse, Wis.
Lockport, N. Y.
Lansing, Mich.
Long Branch, N. J.
Methuen, Mass.
Medford, Mass.
Milford, Mass.
Melrose, Mass.
New Haven, Conn.
New Bedford, Mass.
Natick, Mass.
Norfolk, Va.
New Britian, Conn.
Needham, Mass.
Norwich, Conn.
Oakland, Cal.
Omaha, Neb.
Orange, Mass.
Pittsfield, Mass.
Patterson, N. J.
Port Chester, N. Y.
Providence, R. I.
Portland, Ore.
Pawtucket, R. I.
St. Paul, Minn.
South Framingham, Mass.
Springfield, O.
Spokane, Wash.
Saugus, Mass.
Stratford, Ontario (Canada).
San Francisco, Cal.
Tallapoosa, Ga. (Tallapoosa Water Works Co.)
Taunton, Mass.
Toledo, O.
Traverse, City, Mich.
Tacoma, Wash.
Worcester, Mass.
Woburn, Mass.
Webster, Mass.
Wakefield, Mass.
Winchester, Mass.

FOOTE, PIERSON & CO.

MANUFACTURERS OF

AUTOMATIC AND MANUAL APPARATUS

FOR

FIRE ALARM AND POLICE TELEGRAPH SERVICE

160-162 DUANE STREET NEW YORK CITY

THE LOOMIS
Fire Alarm System

MANUFACTURED BY

The Akron Electrical Mfg. Co.

AKRON, OHIO.

1891.

Gaynor • Rapid • Fire • Alarm • Signal • System

Louisville, Ky. 1889

The Sterling Siren Fire Alarm Co., Inc.

ROCHESTER, N. Y., U. S. A.

Sterling Street Box System

THE Sterling Code Fire Alarm Box is built on the lines of the standard municipal type of box, containing all the desirable features such as lightning arresters, terminal and test block, ground connection and line shunt and telegraph test key.

The operating mechanism is housed in an inner dust-tight, moisture-proof case. The mechanism is of the same high grade self-winding sector type movement as the transmitter, and is especially timed for Siren Coding.

The operating lever is protected by a weatherproof housing with glass front, and is directly connected to the movement without any intermediary lever or parts.

The outside case of the box is made from an aluminum alloy capable of resisting weather conditions for a life time.

One or several boxes may be used and oneCode Control Unit will handle any number of boxes.

McFELL FIRE ALARM SYSTEMS

are noted for lowest maintenance and operation cost ever obtained.

McFell's is a reliable fire alarm system with the highest possible percentage in operation efficiency.

The McFell system provides a circuit which will continue operative, as under normal conditions, while subject to troubles which would suspend service on other systems.

If a tree or a chimney fell upon the wires of the old type system, breaking them, or through some accident to a box on the circuit, if the circuit was open and the current ceased to flow, the protective feature would be entirely lost and the district covered by the said circuit would be without fire protection during such a period of time as the line was undergoing repairs. If this accident were to befall the McFell System no such condition would exist. Each box on the circuit would continue to turn in its proper alarm, as before, with the apparatus gongs and other equipment functioning under normal conditions. It is at once apparent, therefore, that no question or doubt can exist as to the difference in efficiency between the old style loop circuit and the McFell metallic loop circuit, and while we are in a position to furnish the single loop circuit, to those who insist upon this character of apparatus, we do not recommend it, and we would advise strongly against its employment.

Write for our booklet entitled "Fire Alarm Apparatus." It will tell you more about McFell Fire Alarm Systems. Our system operates with wires *short grounded* or *open.*

McFELL SIGNAL CO., Designers, Engineers and manufacturers of the McFell Fire and Police-Alarm Telephone and Telegraph Systems

2857-2859 SOUTH HALSTED STREET, CHICAGO, ILL.

GREGIER FIRE ALARM SYSTEM

The complete Gregier Fire Alarm System is installed in the City of Los Angeles, Cal., (University District).
The City of New York has adopted the Gregier System of a separate pair of wires to each box.

Gregier Signal Company
1114 Stock Exchange Bldg.
Chicago

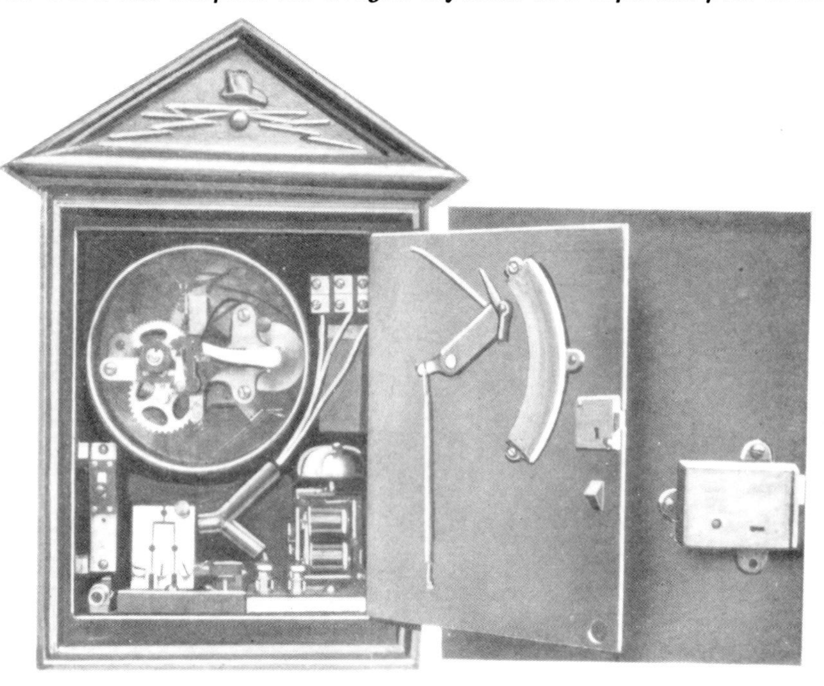

Both Outer and Inner Doors Open.
Type-Four Rounds, Spring Actuated, Key or Keyless Door and Telephone Connections.
These boxes are designed to work in conjunction with the installation of a complete Gregier non-interfering Fire Alarm Telegraph and Telephone System.

Western Electric Company

Chicago New York

LONDON:
Bridge Chambers,
171 Queen Victoria Street, E. C.
and North Woolwich, E.

ANTWERP:
32 Rue Boudewyns

PARIS:
46 Avenue de Breteuil

Police and Fire Alarm Apparatus

Bulletin No. 9002 June, 1900

HARRINGTON SIGN FLASHER CO.

the Former HARRINGTON - SEABERG CORP.

MOLINE, ILLINOIS

HARRINGTON FIRE-ALARM BOXES AS RECOGNIZED BY MUNICIPAL OFFICIALS IN MANY CITIES, HAVE BEEN GIVING SATISFACTORY SERVICE SINCE 1921.

FA-20 FIRE ALARM BOX -:- *Showing Movement*

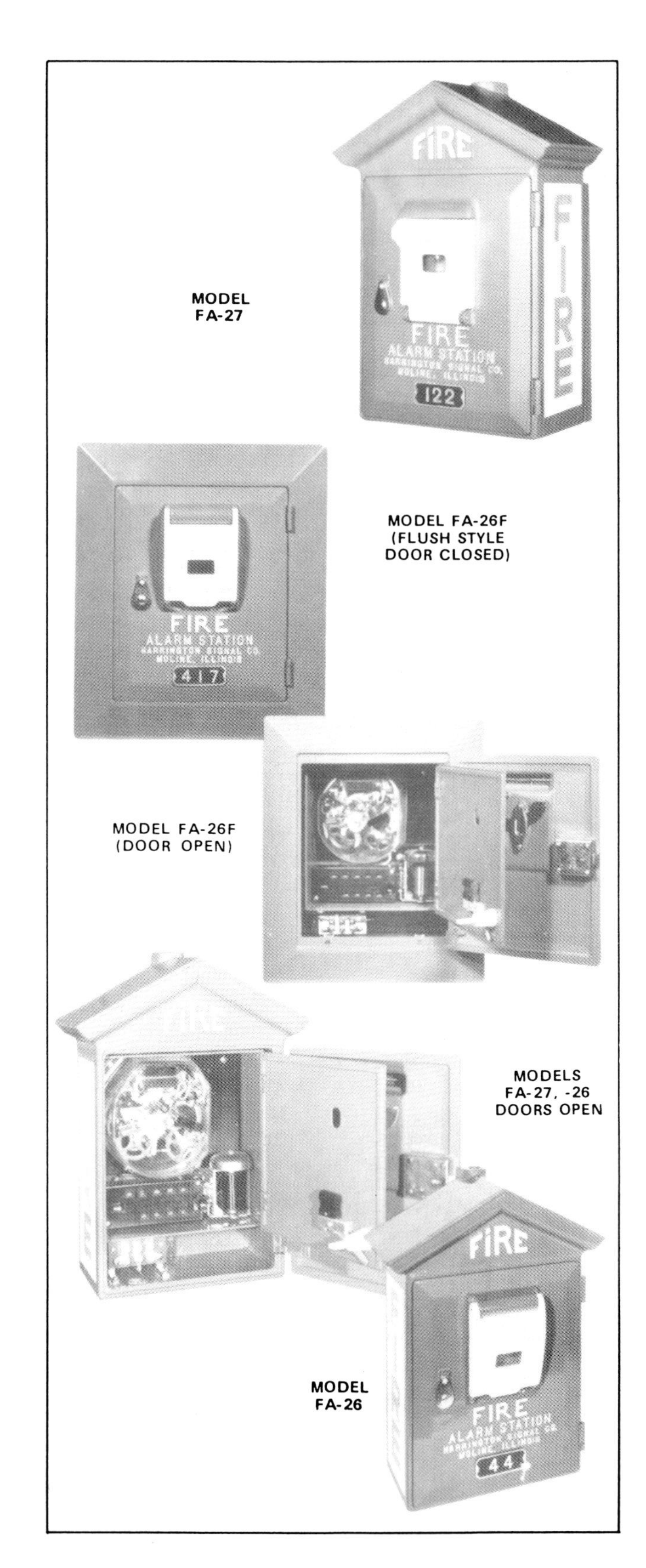

MODEL FA-27

MODEL FA-26F (FLUSH STYLE DOOR CLOSED)

MODEL FA-26F (DOOR OPEN)

MODELS FA-27, -26 DOORS OPEN

MODEL FA-26

GENERAL SIGNAL COMPANY

Manufacturers of
HORNI FIRE ALARM BOXES
CENTRAL STATION EQUIPMENT
115 BROADWAY
NEW YORK 6, N. Y.

Plant
Main & Church Sts.
Fletcher, Ohio

HORNI FIRE ALARM BOXES
will operate in any normal closed metallic fire alarm circuit with other standard types of boxes.

Signal originating at Fire Alarm Box is transmitted to COMPOSITROL and to Register, Gongs, Horns.

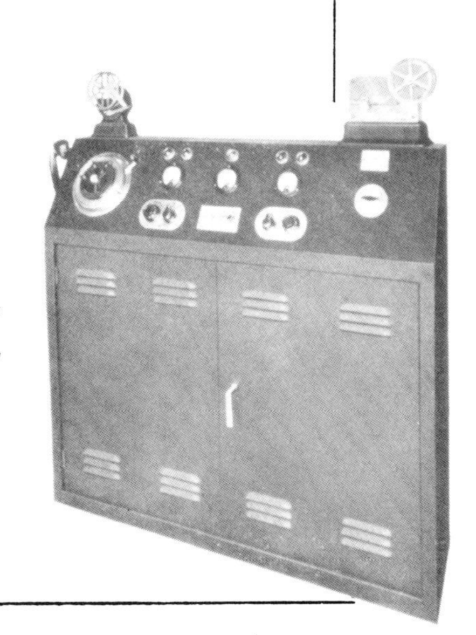

Superior American Fire Alarm & Signal Co.

Formerly General Signal Co.

85 TREMONT STREET

MERIDEN, CONNECTICUT

Plainville, Conn.

~~Fletcher, Ohio~~

CENTRAL OFFICE INSTALLATIONS AS OF 1951

INSTALLED IN:

City of Brattleboro, Vermont

City of East Chicago, Indiana

Township of Cedar Grove, N. J.

Township of Rochelle Park, N. J.

City of Presque Isle, Maine

City of Moncton, New Brunswick

Fire District
Mystic, Connecticut

City of Rensselaer, New York

Village of Rockville Centre, N. Y.

East Chicago, Indiana
(General American Transportation Company Plant)

Village of Elsmere, New York

Village of Woodstock, Vermont

Town of Dover-Foxcroft, Maine

Township of Raritan, New Jersey

Town of North Providence, R. I.

U. S. Government
Ladora, Colorado

Village of Catskill, New York

ON ORDER:

City of Chester, Pennsylvania

Borough of Midland Park, N. J.

Town of Nutley, New Jersey

Borough of Ramsey, New Jersey

City of Portsmouth, Ohio

City of Parkersburg, W. Va.*

State of New York
Brooklyn State Hospital
Brooklyn, New York

City of Springfield, Mass.
(Class A system)

*This will be so constructed that it can be operated either as a Class A or Class B system.

BILLS
FIRE ALARM BOXES

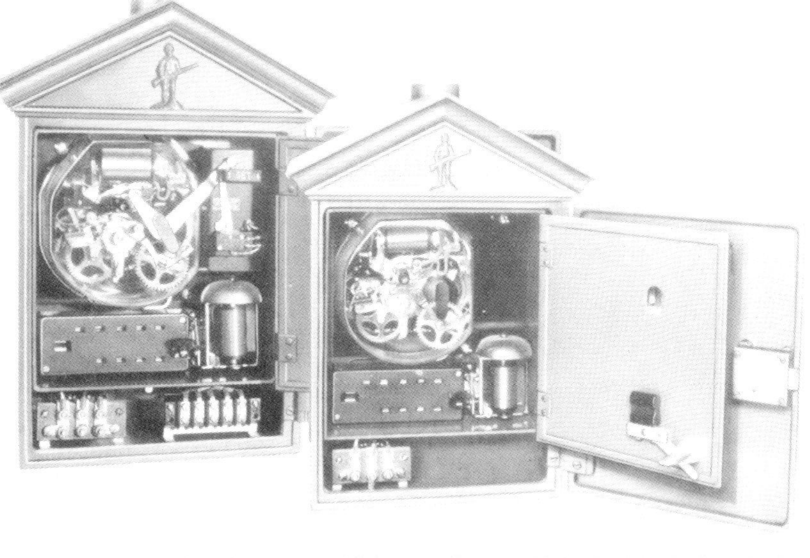

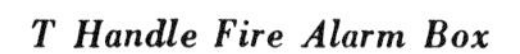

T Handle Fire Alarm Box *Keyless Door Fire Alarm Box*

The above illustration shows two standard fire alarm boxes. One type is known as a "Keyless Door" box. The other is the "T Handle" box.

The Keyless door box has an attachment which loudly rings a bell on the inside of the door when the handle is turned to open the box. This door was devised to call attention to any person opening the box and in this way tends to do away with false alarms. This door is available for any fire alarm box.

The "T Handle" fire alarm box door is less complicated than many later types and is valuable because of its simplicity. A quarter-turn of the handle allows the box to be opened and the alarm lever is in view. Since it is natural for people to open a door in this manner, the simplicity of this door does not further confuse those who are already excited when under the stress of sounding an alarm for fire.

once upon a time...

These Cities Had Other Types of Fire Alarm Equipment

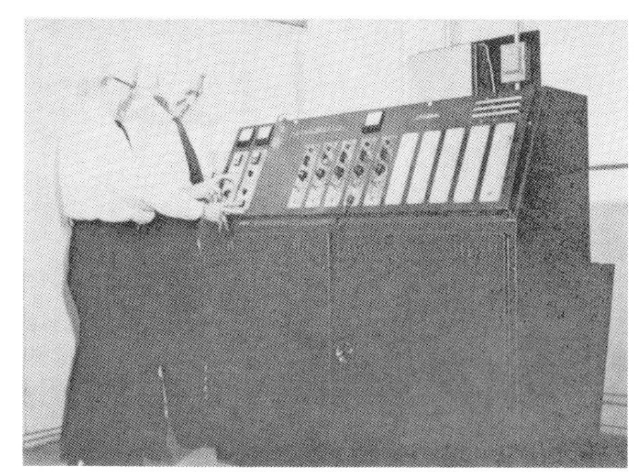

BATAVIA, NEW YORK

ALBANY, GEORGIA

LEXINGTON, MASS.

ELGIN, ILLINOIS

Also—

Mansfield, Mass.; Thompsonville, Conn.; Georgetown, Mass.; Lowell, Mass.; Falmouth, Maine; Webster, Mass.; W. Boylston, Mass.; Cape Elizabeth, Maine; Great Barrington, Mass.; Bedford, Mass.; Braintree, Mass.; Rochester, N. Hampshire; West Bridgewater, Mass.; Hamburg, N.Y.; Chambersburg, Pa.; Montreal, Quebec

B&B ELECTROMATIC CORPORATION
NORWOOD, LOUISIANA 70761
PHONE: 504-629-5261

NEW ENGLAND REPRESENTATIVE:
L. W. BILLS COMPANY
LEXINGTON, MASS. 02173
PHONE: 617-861-0170

Reprinted from
July/

APPENDIX D

A REPRESENTATIVE LIST OF GAMEWELL INSTALLATIONS IN THE UNITED STATES AS OF 1941

ALABAMA

Population

22,345 Anniston
259,678 Birmingham
11,729 Florence
24,042 Gadsden
6,103 Homewood
68,202 Mobile
66,079 Montgomery
18,012 Selma

ARIZONA

8,023 Bisbee
9,828 Douglas
7,157 Globe
3,711 Mesa
7,693 Miami
6,006 Nogales
48,118 Phoenix
5,517 Prescott
32,506 Tucson
4,892 Yuma

ARKANSAS

7,293 Camden
81,679 Little Rock
10,764 Texarkana

CALIFORNIA

35,033 Alameda
8,569 Albany
29,472 Alhambra
3,563 Antioch
5,216 Arcadia — Diaphone Only
1,709 Arcata
2,661 Auburn
26,015 Bakersfield
2,455 Barstow
81,109 Berkeley
17,429 Beverly Hills
13,270 Burlingame
6,299 Calexico
7,961 Chico
2,851 Coalinga
8,014 Colton
2,116 Colusa — Tower Bell Only
1,377 Corning — Tower Bell Only
5,425 Coronado
1,027 Corte Madera
1,824 Crockett
5,669 Culver City
2,623 Delano
2,610 Dunsmuir
8,434 El Centro
3,503 El Segundo — Diaphone Only
3,870 El Cerito
2,336 Emeryville
15,752 Eureka
1,836 Fairfax
3,196 Fontana
52,513 Fresno
3,502 Gilroy
62,736 Glendale
3,817 Grass Valley
7,028 Hanford
5,530 Hayward
3,757 Hollister
2,005 Jackson
1,241 Larkspur
6,788 Lodi
142,032 Long Beach
1,238,048 Los Angeles
3,168 Los Gatos
4,665 Madera
383 Mariposa
6,569 Martinez
5,763 Marysville
7,066 Merced
13,842 Modesto
10,890 Monrovia — Diaphone Only
9,141 Monterey
6,437 Napa
1,701 Nevada City
1,269 Newman — Tower Bell Only
2,203 Newport Beach
2,112 Oakdale — Diaphone Only
284,063 Oakland
13,583 Ontario — Diaphone Only
8,066 Orange
3,698 Oroville
13,652 Palo Alto
76,086 Pasadena
8,245 Petaluma
9,333 Piedmont

Population

9,610 Pittsburgh
20,804 Pomona — Diaphone Only
5,303 Porterville
4,188 Redding
3,517 Red Bluff
8,962 Redwood City
20,093 Richmond
29,696 Riverside
6,425 Roseville
1,355 Ross-Harrington-Seaburg
93,750 Sacramento
10,263 Salinas
4,650 San Anselmo
37,481 San Bernardino
147,995 San Diego
57,651 San Jose
8,276 San Luis Obispo
San Pedro—District of L. A.
8,022 San Rafael
30,332 Santa Ana
33,613 Santa Barbara
6,302 Santa Clara
14,395 Santa Cruz
37,146 Santa Monica
10,636 Santa Rosa
3,667 Sausalito
1,762 Sebastopol
3,047 Selma
2,278 Sonora
6,193 So. San Francisco
47,963 Stockton
1,358 Susanville
3,442 Taft
3,829 Tracy
6,207 Tulare
4,276 Turlock — Diaphone Only
14,476 Vallejo
20,400 Venice — District of L. A.
1,269 Vernon
7,263 Visalia
8,344 Watsonville
4,386 Westwood
2,024 Willows
5,542 Woodland
3,605 Yuba City

COLORADO

705 Aspen
11,223 Boulder
5,938 Canon City
33,237 Colorado Springs
1,427 Cripple Creek
287,261 Denver
5,400 Durango
11,489 Fort Collins
2,426 Golden
12,203 Greeley
1,207 Idaho Springs
3,771 Leadville
1,205 Manitou
707 Ouray
50,096 Pueblo
5,065 Salida
7,195 Sterling
11,732 Trinidad
1,291 Victor

CONNECTICUT

19,898 Ansonia
146,716 Bridgeport
28,451 Bristol
591 Cos Cob
22,261 Danbury
6,951 Darien
4,210 Danielson
1,945 Deep River
10,788 Derby
17,125 East Hartford
2,027 East Port Chester
2,777 Essex
5,981 Greenwich
4,122 Groton
19,020 Hamden
164,072 Hartford
4,436 Jewett City
38,481 Meriden
24,554 Middletown
12,660 Milford
14,315 Naugatuck
68,128 New Britain
162,655 New Haven
29,640 New London
36,019 Norwalk
32,438 Norwich
850 Old Greenwich
7,318 Putnam
7,445 Rockville
2,381 Saybrook
6,890 Seymour
10,113 Shelton

Population

5,125 Southington
8,160 So. Manchester
663 Springdale
46,346 Stamford
2,006 Stonington
8,670 Thompsonville
26,040 Torrington
11,170 Wallingford
1,410 Warehouse Point
99,902 Waterbury
1,037 West Brook
25,808 West Haven
6,073 Westport
12,102 Willimantic
4,073 Windsor Locks
7,883 Winsted

DELAWARE

4,800 Dover
106,597 Wilmington

DISTRICT OF COLUMBIA

486,869 Washington

FLORIDA

5,697 Coral Gables
7,607 Clearwater
1,674 Dania
2,636 De Funiak Springs
3,023 Fernandina
8,666 Fort Lauderdale
9,082 Fort Myers
10,465 Gainesville
2,600 Hialeah
2,869 Hollywood
129,549 Jacksonville
409 Jacksonville Beach
12,831 Key West
18,554 Lakeland
110,637 Miami
6 494 Miami Beach
27,330 Orlando
6,500 Palatka
31,579 Pensacola
6,800 Plant City
3,788 Quincy
40,425 St. Petersburg
8,398 Sarasota
2,912 Sebring
5,597 South Jacksonville
10,700 Tallahassee
101,161 Tampa

GEORGIA

14,507 Albany
8,760 Americus
18,192 Athens
270,367 Atlanta
60,342 Augusta
6,141 Bainbridge
14,022 Brunswick
5,052 Carrollton
43,131 Columbus
8,160 Dalton
6,681 Dublin
4,650 Elberton
53,829 Macon
8,027 Moultrie
6,386 Newman
4,149 Quitman
21,843 Rome
85,024 Savannah
3,996 Statesboro
11,933 Thomasville
13,482 Valdosta
15,510 Waycross

IDAHO

21,544 Boise City
8,297 Coeur d'Alene
9,403 Lewiston
8,206 Nampa
16,471 Pocatello
1,531 Potlatch
3,290 Sand Point
3,634 Wallace

ILLINOIS

866 Algonquin
46,589 Aurora
30,930 Bloomington
1,161 Braidwood
1,461 Carpentersville
3,376,438 Chicago
22,321 Chicago Heights
66,602 Cicero
9,235 Collinsville
36,765 Danville
57,510 Decatur
8,545 DeKalb

Population

3,100 Dundee
74,347 East St. Louis
35,929 Elgin
63,338 Evanston
14,555 Forest Park
22,045 Freeport
3,878 Galena
8,233 Greenville
16,374 Harvey
42,993 Joliet
17,093 Kewanee
10,103 La Grange
2,582 Lemont
1,948 Marengo
4,008 Mendota
32,236 Moline
5,568 Morris
1,445 Mt. Pulaski
63,982 Oak Park
15,094 Ottawa
5,835 Pana
16,129 Pekin
104,969 Peoria
8,272 Pontiac
3,893 Rock Falls
85,864 Rockford
37,953 Rock Island
2,985 Steger
10,012 Sterling
14,728 Streator
2,650 Silvis
4,021 Sycamore
33,499 Waukegan
1,697 West Dundee
3,894 Western Springs
5,471 Woodstock

INDIANA

4,408 Alexandria
39,804 Anderson
9,935 Columbus
12,795 Connersville
10,355 Crawfordsville
54,784 East Chicago
32,949 Elkhart
10,685 Elwood
102,249 Evansville
114,986 Fort Wayne
100,426 Gary
10,397 Goshen
64,560 Hammond
13,420 Huntington
364,161 Indianapolis
5,439 Kendallville
32,843 Kokomo
26,240 Lafayette
15,755 Laporte
18,508 Logansport
6,530 Madison
24,496 Marion
4,962 Martinsville
26,735 Michigan City
28,630 Mishawaka
46,584 Muncie
25,819 New Albany
12,730 Peru
7,505 Princeton
32,493 Richmond
5,709 Rushville
10,618 Shelbyville
104,193 South Bend
62,810 Terre Haute
17,564 Vincennes
8,840 Wabash
9,070 Washington
10,880 Whiting

IOWA

26,755 Burlington
56,097 Cedar Rapids
8,039 Charles City
25,726 Clinton
42,048 Council Bluffs
60,751 Davenport
142,559 Des Moines
41,679 Dubuque
21,895 Fort Dodge
1,241 Garner
15,340 Iowa City
15,106 Keokuk
4,788 Le Mars
17,373 Marshalltown
23,304 Mason City
4,230 Missouri Valley
5,778 Red Oak
3,320 Sheldon
79,183 Sioux City
46,191 Waterloo

KANSAS

14,903 Parsons
64,120 Topeka
111,110 Wichita

KENTUCKY

Population	
65,252	Covington
5,025	Catlettsburg
11,626	Frankfort
4,229	Georgetown
11,668	Henderson
45,736	Lexington
307,746	Louisville
6,485	Ludlow
6,557	Marsville
29,744	Newport
22,765	Owensboro
33,541	Paducah
6,204	Paris
4,033	Shelbyville
8,233	Winchester

LOUISIANA

Population	
23,025	Alexandria
5,121	Bastrop
30,729	Baton Rouge
14,029	Bogulusa
4,003	Bossier City
7,656	Crowley
3,788	Donaldsonville
	Gretna
2,541	Haynesville
2,909	Homer
6,531	Houma
14,635	Lafayette
15,791	Lake Charles
5,623	Minden
26,028	Monroe
5,985	Morgan City
8,603	New Iberia
458,762	New Orleans
6,299	Opelousas
5,024	Plaquemine
76,655	Shreveport
4,442	Thibodaux

MAINE

Population	
18,571	Auburn
17,198	Augusta
28,749	Bangor
4,486	Bar Harbor
9,110	Bath
4,993	Belfast
17,633	Biddeford
2,076	Boothbay Harbor
2,659	Bridgton
6,329	Brewer
6,144	Brunswick
5,470	Calais
3,606	Camden
7,248	Caribou
4,063	Dexter
3,466	Eastport
3,529	Fairfield
1,737	Farmington
3,750	Foxcroft-Dover
2,616	Ft. Fairfield
6,865	Houlton
3,302	Kennebunk
1,024	Kingfield
34,948	Lewiston
3,148	Livermore Falls
2,994	Lubec
1,856	Machias
3,036	Madison
1,837	Mars Hill
2,033	Mechanic Falls
4,767	Mexico
5,830	Millinocket
2,912	Milo
1,731	Newport
612	Northeast Harbor
2,446	Norway
2,664	Oakland
1,620	Old Orchard
7,266	Oldtown
2,075	Pittsfield
70,810	Portland
4,662	Presque Isle
9,075	Rockland
8,726	Rumford Falls
7,233	Saco
13,392	Sanford
6,433	Skowhegan
1,961	So. Paris
13,840	South Portland
2,550	Springvale
4,721	Van Buren
1,971	Washburn
15,454	Waterville
10,807	Westbrook
3,917	Winslow
2,234	Winthrop
1,020	Woodland
2,125	Yarmouth

MARYLAND

Population	
804,874	Baltimore
	Baltimore County
1,650	Bel Air
37,747	Cumberland
30,861	Hagerstown
4,264	Hyattsville

MASSACHUSETTS

Population	
5,872	Abington
12,697	Adams
11,899	Amesbury
5,888	Amherst
9,969	Andover
36,094	Arlington
10,677	Athol
21,769	Attleboro
2,414	Avon
3,060	Ayer
2,603	Bedford
21,748	Belmont
25,086	Beverly
5,880	Billerica
781,188	Boston
15,712	Braintree
9,055	Bridgewater
63,797	Brockton
47,490	Brookline
113,643	Cambridge
5,816	Canton
7,022	Chelmsford
45,816	Chelsea
1,100	Cheshire
43,930	Chicopee
12,817	Clinton
3,083	Cohasset
7,477	Concord
12,957	Danvers
15,136	Dedham
11,323	Easthampton
48,424	Everett
10,951	Fairhaven
15,274	Fall River
4,821	Falmouth
40,692	Fitchburg
5,347	Foxboro
22,210	Framingham
7,028	Franklin
19,399	Gardner
24,204	Gloucester
1,800	Georgetown
5,934	Great Barrington
15,500	Greenfield
2,044	Hamilton
800	Harvard
48,710	Haverhill
6,657	Hingham
3,353	Holbrook
2,846	Holliston
2,973	Hopedale
8,469	Hudson
2,047	Hull
5,599	Ipswich
85,068	Lawrence
4,061	Lee
4,445	Leicester
2,742	Lenox
21,810	Leominster
9,467	Lexington
100,234	Lowell
8,876	Ludlow
103,320	Lynn
1,594	Lynnfield
58,036	Malden
2,636	Manchester
6,364	Mansfield
8,668	Marblehead
1,638	Marion
15,587	Marlboro
7,156	Maynard
59,714	Medford
2,392	Merrimac
21,069	Methuen
8,608	Middleboro
6,957	Milbury
14,741	Milford
1,738	Millis
16,434	Milton
4,918	Monson
1,654	Nahant
3,678	Nantucket
13,589	Natick
10,845	Needham
112,597	New Bedford
15,084	Newburyport
65,276	Newton
22,085	North Adams
24,525	Northampton
6,961	North Andover
10,197	North Attleboro
1,946	Northboro
3,400	North Easton
510	North Scituate
15,049	Norwood
1,333	Oak Bluffs
5,383	Orange
9,577	Palmer
21,345	Peabody
2,922	Pepperell
49,677	Pittsfield
1,606	Plainville
13,042	Plymouth
71,983	Quincy
6,553	Randolph
9,767	Reading
35,680	Revere
7,524	Rockland
3,630	Rockport
43,353	Salem
14,700	Saugus
3,118	Scituate
3,351	Sharon
1,336	Shelburne Falls
2,427	Shirley
103,908	Somerville
14,264	Southbridge
3,476	So. Hadley Falls
6,272	Spencer
149,900	Springfield
10,060	Stoneham
8,204	Stoughton
10,346	Swampscott
37,355	Taunton
1,651	Three Rivers
5,966	Turners Falls
6,285	Uxbridge
1,250	Vineyard Haven
16,318	Wakefield
7,273	Walpole
39,247	Waltham
7,385	Ware
34,913	Watertown
2,937	Wayland
12,992	Webster
11,439	Wellesley
1,119	Wenham
6,409	Westboro
19,775	Westfield
3,332	Weston
16,684	West Springfield
20,882	Weymouth
7,638	Whitman
3,900	Williamstown
4,013	Wilmington
6,202	Winchendon
12,719	Winchester
16,852	Winthrop
19,434	Woburn
195,311	Worcester

MICHIGAN

Population	
13,064	Adrian
6,734	Alma
43,573	Battle Creek
47,355	Bay City
5,571	Berkeley
4,035	Bessemer
4,671	Big Rapids
9,570	Cadillac
1,557	Calumet Township
2,554	Caro
1,888	Caspian
3,377	Clawson
2,995	Crystal Falls
50,358	Dearborn
1,568,662	Detroit
5,550	Dowagiac
13,000	Ecorse
14,524	Escanaba
1,864	Essexville
21,000	Ferndale
156,492	Flint
5,170	Gladstone
8,345	Grand Haven
168,592	Grand Rapids
5,173	Grosse Pointe Vil.
3,533	Grosse Pointe Farms
11,174	Grosse Pointe Park
56,268	Hamtramck
5,795	Hancock
1,892	Harbor Beach
52,959	Highland Park
14,346	Holland
3,757	Houghton
3,615	Howell
7,000	Ionia
11,652	Iron Mountain
9,238	Ishpeming
14,299	Iron Wood
4,665	Iron River
55,187	Jackson
54,786	Kalamazoo
78,397	Lansing
12,336	Lincoln Park
8,898	Ludington
8,087	Manistee
14,789	Marquette
4,053	Melvindale
10,320	Menominee
18,110	Monroe
13,497	Mount Clemens
5,000	Mt. Pleasant
41,390	Muskegon
15,584	Muskegon Hts.
6,552	Negaunee
4,016	Norway
3,245	Otsego
14,496	Owosso
5,740	Petosky
64,928	Pontiac
31,361	Port Huron
1,792	Reed City
17,314	River Rouge
22,904	Royal Oak
80,715	Saginaw
3,389	St. Clair
13,755	Sault Ste. Marie
4,804	South Haven
2,400	Stambaugh
6,950	Sturgis
12,539	Traverse City
4,022	Trenton
3,677	Wakefield
28,368	Wyandotte

MINNESOTA

Population	
1,463	Aurora
12,276	Austin
8,308	Chisholm
6,782	Cloquet
1,243	Coleraine
6,321	Crookston
101,463	Duluth
2,922	East Grand Forks
6,156	Ely
7,484	Eveleth
15,666	Hibbing
2,134	Keewatin
14,038	Mankato
464,356	Minneapolis
7,651	Moorhead
2,555	Nashwauk
7,308	New Ulm
7,654	Owatonna
9,629	Red Wing
20,621	Rochester
21,000	St. Cloud
271,606	St. Paul
7,173	Stillwater
11,963	Virginia
20,850	Winona

MISSISSIPPI

Population	
10,043	Clarksdale
10,743	Columbus
14,807	Greenville
11,123	Greenwood
12,547	Gulfport
18,601	Hattiesburg
48,282	Jackson
31,954	Meridian
22,943	Vicksburg
5,579	Yazoo City

MISSOURI

Population	
16,227	Cape Girardeau
33,454	Joplin
10,491	St. Charles
821,960	St. Louis
57,527	Springfield
25,809	University City

MONTANA

Population	
12,494	Anaconda
16,380	Billings
6,855	Bozeman
39,532	Butte
28,822	Great Falls
6,372	Havre
11,803	Helena
5,358	Lewiston
7,175	Miles City
14,657	Missoula

NEBRASKA

Population	
75,933	Lincoln
214,006	Omaha

NEVADA

Population	
1,596	Carson — Diaphone Only
3,217	Elko
714	Goldfield
18,529	Reno
4,508	Sparks
1,989	Winnemucca — Diaphone Only

NEW HAMPSHIRE

Population	
1,549	Allenstown
1,375	Ashland
20,018	Berlin
1,610	Bristol
25,228	Concord
5,131	Derry
13,573	Dover
4,872	Exeter
6,576	Franklin
3,839	Goffstown
2,000	Gorham
1,319	Greenville
163	Hampton Beach
3,043	Hanover
2,160	Hillsboro
13,794	Keene
12,471	Laconia

Population
2,887 Lancaster
7,073 Lebanon
4,558 Littleton
76,834 Manchester
4,068 Milford
31,463 Nashua
2,511 Newmarket
4,659 Newport
1,414 North Walpole
2,521 Peterboro
2,018 Pittsfield
2,470 Plymouth
14.495 Portsmouth
10,209 Rochester
2.751 Salem
5,680 Somersworth
1,040 Sunapee
1,428 Suncook
1,712 Tilton
1,267 Troy
2,358 Wolfeboro
1,230 Woodsville

NEW JERSEY

14,981 Asbury Park
66,198 Atlantic City
88,979 Bayonne
26,974 Belleville
3,491 Belmar
2,073 Belvidere
38,077 Bloomfield
7,372 Bound Brook
3,306 Bradley Beach
15,699 Bridgeton
10,844 Burlington
3,392 Butler
5,144 Caldwell
118,700 Camden
2,637 Cape May City
5,425 Carlstadt
46,875 Clifton
11,103 Cranford
800 Deal
10,031 Dover
2,686 East Newark
68,020 East Orange
7,080 East Rutherford
4,089 Edgewater
114,589 Elizabeth
17,805 Englewood
2,260 Fair Haven
6,894 Freehold
3,334 Garwood
7,365 Glen Ridge
13,796 Gloucester
24,568 Hackensack
4,812 Haledon
15,601 Harrison
5,658 Hasbrouck Hts.
8,691 Highland Park
56,261 Hoboken
Holly Beach
56,733 Irvington
316,715 Jersey City
40,716 Kearney
4,940 Keyport
3,060 Lakewood
21,206 Linden
4,155 Little Ferry
18,399 Long Branch
17,356 Lyndhurst
21,510 Maplewood
2,913 Margate City
8,548 Millburn
14,705 Millville
42,017 Montclair
4,896 Moorestown
15,197 Morristown
2,258 Neptune City
Neptune Township
442,337 Newark
34,555 New Brunswick
5,401 Newton
8,263 North Arlington
40,200 North Bergen
9,760 North Plainfield
20,572 Nutley
5,525 Ocean City
3,060 Ocean Grove
35,399 Orange
9,065 Palisades Park
62,959 Passaic
138,513 Paterson
43,516 Perth Amboy
19,255 Phillipsburg
5,867 Piscatawaytown
34,422 Plainfield
3,105 Pompton Lakes
16,011 Rahway
4,751 Raritan
11,622 Red Bank
4,671 Ridgefield
10,764 Ridgefield Park
12,188 Ridgewood
13,021 Roselle
14,915 Rutherford
8,255 Somerville
8,476 South Amboy

NEW MEXICO

Population
13,630 South Orange
10,759 South River
14,556 Summit
123,356 Trenton
58,659 Union City
6,674 Ventnor City
7,161 Verona
7,556 Vineland
14,775 Weehawken
1,686 West Long Branch
37,107 West New York
24,327 West Orange
5,330 Wildwood
25,266 Woodbridge Township
8,172 Woodbury
26,570 Albuquerque
2,518 Clayton
8,027 Clovis
3,377 Deming
4,719 Las Vegas City
4,378 Las Vegas Town

NEW YORK

1,538 Addison
1,952 Alexandria Bay
127,412 Albany
4,878 Albion
34,817 Amsterdam
36,652 Auburn
3,845 Baldwinsville
17,375 Batavia
11,933 Beacon
1,202 Bellerose Terrace, L. I.
76,662 Binghamton
3,511 Brockport
573,076 Buffalo
3,000 Canajahorie
7,541 Canandaigua
4,235 Canastota
2,822 Canton
4,460 Carthage
5,082 Catskill
1,788 Cazenovia
5,065 Cedarhurst
20,849 Cheektowaga
2,374 Clyde
23,226 Cohoes
15,777 Corning
15,043 Cortland
2,195 Coxsackie
400 Delmar
6,536 Depew
5,741 Dobbs Ferry
3,309 Dolgeville
17,082 Dunkirk
4,340 East Rockaway
4,646 East Syracuse
510 Eggertsville
47,397 Elmira
492 Elmont
2,935 Elmsford
Elmwood
16,231 Endicott
3,579 Falconer
10,016 Floral Park
2,725 Fort Plain
5,814 Fredonia
15,476 Freeport
12,462 Fulton
16,053 Geneva
714 Gardenville
11,430 Glen Cove
18,531 Glens Falls
23,099 Gloversville
2,891 Goshen
4,015 Gouverneur
3,062 Greenport
2,447 Groton-on-Hudson
1,427 Hancock
9,215 Harrison
5,621 Haverstraw
12,650 Hempstead
10,446 Herkimer
6,722 Hicksville
3,195 Homer
16,250 Hornell
12,337 Hudson
6,449 Hudson Falls
9,890 Ilion
3,067 Irvington
20,708 Ithaca
45,155 Jamestown
13,567 Johnson City
10,801 Johnstown
28,088 Kingston
23,948 Lackawanna
7,040 Lancaster
4,474 LeRoy
5,282 Larchmont
3,427 Liberty
11,105 Little Falls
2,244 Liverpool
23,160 Lockport
11,993 Lynbrook
3,956 Lyons
8,657 Malone

Population
11,766 Mamaroneck
10,687 Massena
7,924 Mechanicville
6,071 Medina
21,276 Middletown
8,155 Mineola
2,835 Mohawk
1,621 Monroe
5,127 Mount Kisco
60,449 Mount Vernon
7,649 Newark
31,275 Newburgh
1,885 New Hartford
3,314 New Hyde Park
54,000 New Rochelle
4,006 New York Mills
75,460 Niagara Falls
7,417 North Tarrytown
19,019 North Tonawanda
8,378 Norwich
1,880 Norwood
16,915 Ogdensburg
21,790 Olean
10,588 Oneida
12,536 Oneonta
15,241 Ossining
22,652 Oswego
4,742 Owego
2,592 Palmyra
6,860 Patchogue
17,125 Peekskill
2,053 Pelham
4,908 Pelham Manor
5,329 Penn Yan
4,231 Perry
13,349 Plattsburg
4,540 Pleasantville
22,662 Port Chester
10,242 Port Jervis
3,060 Port Washington, L. I.
40,288 Poughkeepsie
4,136 Potsdam
996 Red Hook
1,569 Rhinebeck
11,223 Rensselaer
1,333 Richfield Springs
328,132 Rochester
13,718 Rockville Centre
32,338 Rome
8,712 Rye
2,273 St. Johnsville
9.577 Salamanca
8,020 Saranac Lakes
13,169 Saratoga
9,690 Scarsdale
95,692 Schenectady
7,437 Scotia
6,443 Seneca Falls
1,882 Skaneateles
7,986 Solvay
2,689 South Glens Falls
3,948 Spring Valley
3,757 Suffern
209,326 Syracuse
6,841 Tarrytown
12,681 Tonawanda
72,763 Troy
6,138 Tuckahoe
2,040 Tuxedo Park
101,740 Utica
11,790 Valley Stream
4,283 Walden
3,477 Warsaw
2,443 Warwick
4,047 Waterloo
32,205 Watertown
16,083 Watervliet
2,956 Watkins
5,662 Waverly
5,674 Wellsville
West Seneca
13,000 West Hempstead
3,375 Whitesboro
5,191 Whitehall
35,830 White Plains
134,646 Yonkers

NORTH CAROLINA

50,193 Asheville
2,957 Beaufort
737 Black Mountain
9,737 Burlington
5,117 Canton
2,699 Chapel Hill
82,675 Charlotte
52,037 Durham
10,037 Elizabeth City
2,056 Farmville
17,093 Gastonia
14,985 Goldsboro
2,972 Graham
53,569 Greensboro
9,194 Greenville
6,345 Henderson
7,363 Hickory
36,745 High Point
11,362 Kinston
3,312 Laurinburg

Population
6,532 Lenoir
9,652 Lexington
4,140 Lumberton
5,619 Mooresville
3,483 Morehead City
2,254 Mt. Holly-Codewell
11,981 New Bern
4,101 Oxford
714 Pinehurst
37,379 Raleigh
2,906 Rockingham
21,412 Rocky Mount
16,951 Salisbury
4,253 Sanford
10,491 Statesville
1,378 Sullivans Island
Tarboro
10,090 Thomasville
1,536 Wake Forest
7,035 Washington
980 Wendell
32,270 Wilmington
12,613 Wilson
75,274 Winston-Salem

NORTH DAKOTA

28,619 Fargo
17,112 Grand Forks

OHIO

255,040 Akron
23,047 Alliance
23,934 Barberton
13,327 Bellaire
9,543 Bellefontaine
10,027 Bucyrus
15,000 Campbell
104,906 Canton
18,340 Chillicothe
451,160 Cincinnati
7,369 Circleville
900,429 Cleveland
290,564 Columbus
9,691 Conneaut
10,908 Coshocton
1,807 Covington
200,982 Dayton
8,818 Defiance
9,716 Dover
39,667 East Cleveland
23,329 East Liverpool
19,363 Findlay
7,674 Galion
7,036 Greenville
52,176 Hamilton
4,040 Hillsboro
16,621 Ironton
7,069 Kenton
70,509 Lakewood
18,716 Lancaster
42,287 Lima
44,512 Lorain
33,525 Mansfield
Mariemont
14,285 Marietta
31,084 Marion
14,524 Martin's Ferry
26,400 Massillon
1,714 McDonald
29,992 Middletown
31,000 Newark
5,931 New Boston
33,411 Norwood
16,009 Piqua
3,563 Pomeroy
42,560 Portsmouth
7,487 St. Bernard
10,622 Salem
24,622 Sandusky
68,743 Springfield
35,422 Steubenville
11,249 Struthers
16,428 Tiffin
290,718 Toledo
8,675 Troy
8,472 Van Wert
41,062 Warren
7,956 Wellsville
10,742 Wooster
3,767 Wyoming
170,002 Youngstown
36,440 Zanesville

OKLAHOMA

15,741 Ardmore
14,763 Bartlesville
26,399 Enid
7,694 Henryetta
11,804 McAlester
32,026 Muskogee
185,389 Oklahoma City
17,097 Okmulgee
10,533 Sapulpa
141,258 Tulsa

Population

OREGON

5,325 Albany
4,544 Ashland
10,349 Astoria
7,858 Baker — Diaphone Only
2,757 Hood River — Tower Bell Only
16,093 Klamath Falls
8,050 La Grande
4,012 North Bend
5,761 Oregon City
6,621 Pendleton
301,815 Portland
5,883 The Dalles

PENNSYLVANIA

27,116 Aliquippa
92,563 Allentown
82,054 Altoona
20,227 Ambridge
9,587 Archbald
4,263 Aspinwall
5,824 Bangor
17,147 Beaver Falls
12,660 Berwick
57,892 Bethlehem
8,260 Blakeley
9,093 Bloomsburg
6,250 Brackenridge
19,329 Braddock
19,306 Bradford
11,799 Bristol
2,869 Brownsville
12,558 Canonsburg
20,061 Carbondale
12,596 Carlisle
13,788 Chambersburg
11,260 Charleroi
59,164 Chester
15,291 Clairton
9,221 Clearfield
6,921 Coaldale
14,582 Coatesville
13,290 Connellsville
7,152 Corry
12,395 Dickson City
13,190 Dormont
22,627 Dunmore
21,396 Duquesne
34,468 Easton
6,214 East Pittsburgh
6,099 East Stroudsburg
3,063 Ebensburg
4,821 Edgewood
8,847 Edwardsville
3,940 Elizabethtown
12,323 Ellwood City
115,967 Erie
7,493 Etna
6,127 Ford City
6,224 Forty Fort
4,568 Fountain Hill
8,034 Frackville
10,254 Franklin
7,098 Freeland
3,458 Gallitzen
16,508 Greensburg
11,805 Hanover
80,339 Harrisburg
36,765 Hazelton
20,141 Homestead
5,490 Honesdale
2,252 Hughestown
15,126 Jeannette
4,692 Jessup
66,993 Johnstown
6,232 Kane
3,091 Kennet Sq.
21,600 Kingston
7,808 Kittanning
2,841 Kutztown
59,949 Lancaster
9,632 Lansford
9,322 Larksville
25,561 Lebanon
4,171 Lemoyne
3,308 Lewisburg
14,784 Mahanoy City
3,281 McDonald
54,632 McKeesport
16,698 Meadville
1,959 Mifflinburg
8,166 Millvale
9,392 Minersville
8,675 Monongahela City
17,967 Mt. Carmel
5,869 Mt. Pleasant
26,043 Nanticoke
5,916 Natrona
5,505 Nazareth
48,674 New Castle
16,762 New Kensington
35,853 Norristown
9,839 Northampton
16,782 North Braddock
3,670 North East

Population

22,075 Oil City
12,661 Old Forge
10,743 Olyphant
5,628 Parsons
3,993 Peckville
4,310 Pen Argyl
1,950,961 Philadelphia
3,600 Philipsburg
12,029 Phoenixville
6,317 Pitcairn
669,817 Pittsburgh
18,246 Pittston
1,224 Plains
306 Pleasant Valley
16,543 Plymouth
19,430 Pottstown
24,300 Pottsville
7,956 Rankin
111,171 Reading
6,313 Ridgeway
7,726 Rochester
6,714 Scottdale
20,274 Shamokin
25,908 Sharon
8,642 Sharpsburg
21,782 Shenandoah
2,451 Shickshinny
3,857 Souderton
3,227 South Fork
2,520 South Greensburgh
3,105 Southwest Greensburgh
1,236 Spring Grove
7,296 St. Clair
7,433 St. Marys
13,291 Steelton
5,961 Stroudsburg
5,567 Summit Hill
15,626 Sunbury
3,203 Susquehanna
16,029 Swissvale
12,936 Tamaqua
9,511 Tarentum
10,428 Taylor
8,027 Throop
8,055 Titusville
4,104 Towanda
10,690 Turtle Creek
3,788 Union City
19,545 Uniontown
11,479 Vandergrift
14,836 Warren
24,545 Washington
12,325 West Chester
3,552 West Homestead
2,953 West Newton
7,940 West Pittston
4,908 West Reading
6,028 Westview
86,626 Wilkes-Barre
29,539 Wilkinsburg
45,729 Williamsport
9,205 Windber
55,254 York

RHODE ISLAND

2,754 Anthony
11,953 Bristol
7,677 Burrillsville
25,898 Central Falls
143,433 Scranton
5,599 Sewickley
6,430 Coventry
42,911 Cranston
1,000 Cumberland
29,995 East Providence
2,805 Harrisville
1,599 Jamestown
985 Lakewood
4,590 Manville
1,384 Narragansett Pier
5,488 Natick
27,612 Newport
1,000 North Scituate
2,805 Pascoag
77,149 Pawtucket
3,366 Phoenix
252,981 Providence
4,580 River Point
1,416 Saylesville
3,366 Valley Falls
7,974 Warren
23,196 Warwick
10,995 Westerly
49,376 Woonsocket

SOUTH CAROLINA

14,383 Anderson
5,183 Camden
62,265 Charleston
51,581 Columbia
5,556 Darlington
14,774 Florence
5,082 Georgetown
11,020 Greenwood

Population

5,069 Hartsville
7,298 Newberry
8,776 Orangeburg
11,322 Rock Hill
28,723 Spartanburg
11,780 Sumter
7,419 Union

SOUTH DAKOTA

5,733 Lead
10,942 Mitchell
33,362 Sioux Falls

TENNESSEE

119,798 Chattanooga
9,136 Cleveland
7,882 Columbia
4,588 Harriman
105,802 Knoxville
253,143 Memphis
153,866 Nashville

TEXAS

23,175 Abilene
43,132 Amarillo
57,732 Beaumont
7,569 Breckenridge
22,021 Brownsville
27,741 Corpus Christi
260,475 Dallas
11,693 Del Rio
13,850 Denison
4,821 Edinburg
102,421 El Paso
163,447 Fort Worth
52,938 Galveston
292,352 Houston
5,036 Longview
7,311 Lufkin
5,338 Marlin
7,913 Orange
50,092 Port Arthur
25,308 San Angelo
231,542 San Antonio
15,713 Sherman
15,345 Temple
16,602 Texarkana
4,421 Victoria
52,848 Waco
43,690 Wichita Falls

UTAH

40,272 Ogden
14,766 Provo
140,267 Salt Lake City

VERMONT

11,307 Barre
1,363 Barton
3,930 Bellows Falls
7,390 Bennington
2,891 Brandon
8,709 Brattleboro
1,190 Bristol
24,789 Burlington
2,165 Derby
1,621 Essex Junction
2,289 Fair Haven
1,667 Hardwick
1,642 Ludlow
2,003 Middlebury
7,837 Montpelier
1,822 Morrisville
5,094 Newport
1,301 Orleans
1,570 Poultney
2,515 Proctor
17,315 Rutland
8,020 St. Albans
7,920 St. Johnsbury
4,943 Springfield
West Derby
2,590 White River Junction
5,308 Winooski
1,312 Woodstock

VIRGINIA

24,149 Alexandria
15,245 Charlottesville
6,839 Clifton Forge
6,538 Covington
22,247 Danville
6,819 Fredericksburg
11,327 Hopewell
40,661 Lynchburg
34,417 Newport News
129,710 Norfolk
28,564 Petersburg

Population

45,704 Portsmouth
182,929 Richmond
69,206 Roanoke
Schoolfield
4,841 South Boston
11,990 Staunton
10,271 Suffolk
3,327 Wytheville

WASHINGTON

21,723 Aberdeen
30,823 Bellingham
10,170 Bremerton
8,058 Centralia
4,907 Chehalis
4,621 Ellensberg
30,567 Everett
12,766 Hoquiam
11,733 Olympia
3,496 Pasco
3,979 Port Townsend
3,828 Raymond
365,583 Seattle
115,514 Spokane
106,817 Tacoma
15,996 Walla-Walla
11,627 Wenatchee
22,101 Yakima

WEST VIRGINIA

3,950 Benwood
19,339 Bluefield
60,408 Charleston
75,572 Huntington
14,411 Moundsville
5,100 Nitro
29,623 Parkersburg
5,720 Richwood
3,072 Sistersville
8,572 Weirton
61,659 Wheeling

WISCONSIN

8,610 Antigo
25,267 Appleton
10,622 Ashland
5,545 Baraboo
23,611 Beloit
4,114 Burlington
9,539 Chippewa Falls
26,287 Eau Claire
1,124 Ellsworth
5,793 Ft. Atkinson
26,449 Fond du Lac
37,415 Green Bay
3,264 Hurley
21,628 Janesville
50,262 Kenosha
39,614 La Crosse
3,073 Lake Geneva
57,899 Madison
22,963 Manitowoc
13,734 Marinette
578,249 Milwaukee
5,015 Monroe
5,030 Oconto
40,108 Oshkosh
67,542 Racine
8,019 Rhinelander
39,251 Sheboygan
10,706 South Milwaukee
36,113 Superior
2,792 Viroqua
10,613 Watertown
23,758 Wausau
21,194 Wauwatosa
34,671 West Allis

WYOMING

1,749 Buffalo
16,619 Casper
17,361 Cheyenne
3,075 Evanston
8,609 Laramie
4,868 Rawlins
8,440 Rock Springs
8,536 Sheridan
751 So. Superior

ALASKA

973 Dawson
4,043 Juneau

HAWAII

137,582 Honolulu

A REPRESENTATIVE LIST OF GAMEWELL INSTALLATIONS IN FOREIGN COUNTRIES

Population

ARGENTINA

- 2,250,000 Buenos Aires
- 485,000 Rosario

BRAZIL

- 1,150,000 São Paulo

BRITISH EMPIRE

UNITED KINGDOM

- Aldershot
- 415,000 Belfast
- 102,000 Blackpool
- 301,000 Bradford
- 411,000 Bristol
- 224,000 Cardiff
- Chigwell
- Chingford
- 240,000 Croydon
- Dagenham
- Dorking
- 77,000 Edmonton
- 27,000 Epsom
- Eton
- East Barnet
- 147,000 East Ham
- 68,000 Enfield
- Esher
- 26,000 Harrow
- 56,000 Hendon
- 55,000 Hove
- 863,000 Liverpool
- 4,300,000 London
- 758,000 Manchester
- Manchester Ship Canal
- Merton
- Northwood
- 34,000 Ramsgate
- 38,000 Richmond
- 24,000 Rugby
- Salford
- 120,000 Southend
- Stretford
- Surbiton
- Sutton and Cheam
- 186,000 Sunderland
- Wealdstone
- Whitley Bay
- 184,000 Willisden
- 59,000 Wimbledon
- 54,000 Woodgreen
- 46,000 Worthing
- Woolwich Arsenal
- Windsor Castle
- Hampton Court Royal Palaces
- H. M. Factory, Irvine
- Morris Motors, Ltd.
- Siemens Bros., Ltd.

BRITISH MALAYA

- Ipoh
- 101,000 Penang
- 600,000 Singapore

BRITISH WEST INDIES

- 72,000 Kingston, Jamaica

Dominion of Canada

ALBERTA

- 83,304 Calgary
- 85,676 Edmonton
- 13,520 Lethbridge
- 9,590 Medicine Hat

BRITISH COLUMBIA

- Essondale
- 2,732 Fernie
- 6,167 Kamloops
- 6,745 Nanaimo
- 5,992 Nelson
- 17,524 New Westminster
- Point Grey
- Powell River
- 6,350 Prince Rupert
- 2,848 Rossland
- 3,000 Revelstoke
- South Vancouver
- Tadanac
- 7,573 Trail
- 246,593 Vancouver
- 3,937 Vernon
- 39,082 Victoria
- Warfield

MANITOBA

- 16,388 Brandon
- 6,537 Portage la Prairie
- 16,255 St. Boniface
- 5,576 Transcona
- 215,602 Winnipeg

NEW BRUNSWICK

- 3,300 Bathurst
- 6,505 Campbellton
- 4,017 Chatham
- 6,430 Edmundston
- 3,500 Fairville
- 8,830 Fredericton
- 20,689 Moncton
- 2,234 Sackville
- 47,514 St. John
- 3,437 St. Stephen
- 2,252 Sussex
- 3,259 Woodstock
- West St. John

NEWFOUNDLAND

- 42,645 St. Johns

NOVA SCOTIA

- 7,450 Amherst
- 3,262 Bridgewater
- 9,100 Dartmouth
- 20,706 Glace Bay
- 59,275 Halifax
- 2,900 Inverness
- 3,033 Kentville
- 8,858 New Glasgow
- 6,639 North Sydney
- 5,002 Stellarton
- 23,089 Sydney
- 2,613 Trenton
- 7,901 Truro
- 7,000 Yarmouth

ONTARIO

- 2,006 Alexandria
- 2,759 Amherstburg.
- Ansonville
- 2,587 Aurora
- 7,776 Barrie
- 4,080 Bowmanville
- 5,532 Brampton
- 30,107 Brantford
- 9,736 Brockville
- 4,105 Carleton Place
- 2,000 Chapleau
- 14,569 Chatham
- 3,885 Cobalt
- 11,126 Cornwall
- 661 Crystal Beach
- 1,476 Deseronto
- 5,026 Dundas
- 14,251 East Windsor
- 3,000 Fergus
- 26,277 Fort William
- 3,592 Ganonoque
- 155,547 Hamilton
- 5,177 Hawkesbury
- 1,476 Iroquois Falls
- 6,766 Kenora
- 23,439 Kingston
- Kirkland Lake
- 30,793 Kitchener
- 7,505 Lindsay
- 71,148 London
- 2,523 Merriton
- 3,497 Napanee
- 3,748 New Market
- 19,046 Niagara Falls
- 15,528 North Bay
- North Toronto
- 3,857 Oakville
- 2,614 Orangeville
- 8,183 Orillia
- 23,439 Oshawa
- 126,872 Ottawa
- 12,839 Owen Sound
- 4,137 Paris
- 9,368 Pembroke
- 22,327 Peterboro
- 3,580 Picton
- Point Edward
- 19,818 Port Arthur
- 3,000 Portsmouth
- 2,984 Prescott
- 10,715 Sandwich
- 18,191 Sarnia
- 23,082 Sault Ste. Marie
- Scarboro
- Schumacher
- 1,686 Seaforth
- 5,226 Simcoe
- 7,108 Smith Falls
- 1,000 South Porcupine
- Steelton
- 17,742 Stratford
- 24,743 St. Catherines
- 15,430 St. Thomas
- 18,518 Sudbury
- 5,092 Thorold
- 14,200 Timmins
- 631,207 Toronto
- 10,105 Walkerville
- 8,095 Waterloo
- 63,108 Windsor
- 11,000 Woodstock
- 10,709 Welland

QUEBEC

- 3,729 Beauharnais
- 8,748 Cap de la Madeleine
- 955 Chambly Canton
- 11,877 Chicoutimi
- 6,609 Drummondville
- 4,000 East Angus
- 4,205 Farnham
- Frazerville
- 10,587 Granby
- 6,461 Grandmère
- 29,433 Hull
- 2,778 Iberville
- Isle Maligne
- 10,765 Joliette
- 4,500 Kenogami
- 18,630 Lachine
- 7,871 La Tuque
- Longue Point
- 5,407 Longueuil
- Maisonneuve
- 818,577 Montreal
- 2,242 Montreal East
- 28,641 Outremont
- 3,000 Plessisville
- 2,970 Pointe-aux-Trembles
- 2,342 Port Alfred
- 130,594 Quebec
- 2,596 Richmond
- 8,499 Rivière de Loup
- 15,345 Shawinigan Falls
- 28,933 Sherbrooke
- 2,417 Ste. Anne de Bellevue
- 8,967 St. Jerome
- 3,970 St. Joseph d'Alma
- 13,448 St. Hyacinthe
- 11,256 St. Johns
- 6,075 St. Lambert
- 5,348 St. Laurent
- 4,185 St. Pierre Aux Liens
- 10,701 Thetford Mines
- 35,450 Three Rivers
- 11,411 Valleyfield
- 60,745 Verdun
- 6,213 Victoriaville
- 24,235 Westmount

SASKATCHEWAN

- 19,782 Moose Jaw
- 4,727 North Battleford
- 11,050 Prince Albert
- 53,389 Regina
- 41,606 Saskatoon
- 5,065 Swift Current
- 5,325 Weyburn

INDIA

- 290,000 Colombo
- Madras Railway, Calcutta
- Ludlow Jute Mills, Chengail
- 400,000 Rangoon

NEW ZEALAND

(Gamewell-Duplex Systems)

- 218,000 Auckland
- Castlecliff
- 129,000 Christchurch
- Devonport
- 86,000 Dunedin
- Dannevirke
- 16,000 Gisborne
- 18,000 Hamilton
- 17,000 Hastings
- 5,000 Hawera
- 24,000 Invercargill
- Mt. Albert
- Mt. Eden
- 19,000 Napier
- 18,000 New Plymouth
- 11,000 Onehunga
- 8,000 Oamaru
- Parnell
- 10,000 Petone
- Remuera
- Sydenham
- St. Albans
- St. Kilda
- Suva
- Tamaki
- Takapuna
- 18,000 Timaru
- 28,000 Wanganui
- Woolston

UNION OF SOUTH AFRICA

- 60,000 Bloemfontein
- 306,000 Capetown
- 55,000 East London
- 52,000 Germiston
- 400,000 Johannesburg
- 40,000 Kimberley
- 110,000 Pretoria
- General Hospital, Johannesburg
- South African Railway & Harbour, Durban
- South African Railway, Port Elizabeth

. . . .

GERMANY

- 323,000 Bochum
- 500,000 Duisburg
- 200,000 Gladbach
- 450,000 Hannover
- 128,000 Kattowitz
- Linden
- Ruhrort
- Schoneburg

PHILIPPINE ISLANDS

- 65,000 Cebu
- 300,000 Manila

REPUBLIC OF PANAMA

- 40,000 Colon
- 125,000 Panama City
- Colon Hospital
- Santo Tomas Hospital
- United Fruit Company, Almirante

RUSSIA

- 1,700,000 Leningrad
- Singer Sewing Machine Company, Podolsk

ACKNOWLEDGEMENTS

Would-be authors are often lulled into a false sense of ease whenever they undertake a book they hope will be definitive. Surely, I thought, all the research and illustrations needed would quickly be found at the headquarters of The Gamewell Company. While my contact there was intrigued by my goal of producing the complete story of Gamewell, corporate files were all but bereft of material.

My quest turned to the Smithsonian Institution where I was again encouraged in my project by a curator who remarked that historians had long hoped for a book which told the story of the little red fire alarm boxes from their earliest beginnings to the days when these commonplace sights of Americana began to vanish in favor of improved telephone services, including the 9-1-1 system, radio, microwave, computers and even satellites for reporting fires and quickly alerting firefighters.

As my search broadened, I was fortunate to know colleagues and others who had assisted me with earlier books on fire service history. New friends and eager help came from many, including collectors and dealers in fire alarm paraphernalia, plus former employees of Gamewell as well as the National Board of Fire Underwriters, whose standards, starting in 1904, became a virtual blueprint for fire alarm systems throughout the United States. Fortunately their memories and recollections were as vivid as was the value of their collections which they freely permitted me to use.

Notable among those extremely knowledgeable sources was John Bryan, Jr., of Towson, Maryland, a Baltimore County Fire Department Fire Protection Engineer. John not only welcomed my publisher, Fred Conway (himself an avid collector of fire alarm memorabilia) and me to his home where we saw his collection, but spent many hours schooling me in fire alarm technology and history while freely making his superb collection of alarm boxes available for photographing by Dan Thacker, also of the Baltimore County Fire Department.

Robert W. Fitz, retired Deputy Chief of the Lebanon, New Hampshire, Fire Department, packaged and sent me his entire collection of photographs, illustrations, Gamewell and other manufacturers' sales catalogs, plus invaluable documents which otherwise would not have been available to me.

Nor can I overly express my gratitude, not to mention admiration, for the assistance of Division Chief Jack Supple of the Buffalo, New York, Fire Department. Although Hoodoo Box 29 was no stranger to me — I first wrote about it when I was a **Buffalo Evening News** reporter — Chief Supple was always available and eager to help flush out, verify and doublecheck my information; especially details of the major fire that Chief Supple commanded at the last major blaze at the Hoodoo Box.

Among the joys of authorship are the surprises that await. Totally lacking in insightful biographical material on John Gamewell, I played a hunch and called the Director of the Camden, South Carolina, Archives and Museum. Perhaps the Museum would have necessary data on Gamewell, a native South Carolinian, who lived many years in Camden. Director Risher R. Fairey interrupted me when I introduced myself. The Director said I was well-known to the Museum because my 1976 Pulitzer Prize candidate book, **Fire Engines, Firefighters,** was their main source of information on Gamewell. Delving deeper and with the additional help of Dr. Charles H. Lesser of the South Carolina Department of Archives & History, significant genealogical and other material emerged from obscurity and was utilized in this book.

I must, however, reserve my deepest gratitude of all to Dominic Salvatore, whose editing and publishing experience over a span of 25 years with major New York publishing houses, enabled me to complete the book when many factors militated against it. His suggestions, editing, word processing assistance, and ongoing encouragement kept me at the typewriter when I was ready to quit, light up a cigar and enjoy Mozart which would have enabled me to have been as far removed from this book as was tantalizingly possible.

Dominic should author a book, "How To Keep Authors Happily Inspired and Perspired." It would be a best-seller. I know of no way to adequately show my gratitude to Dominic while, at the same time, apologizing to his bride, Carol, for stealing so much of Dominic's freely-given time, editing and publishing expertise.

I am deeply grateful, too, for those many fire alarm authorities whose encouragement and knowledge gave me insights which I would not otherwise have had. They include:

Chief George "Smokey" Bass, Retd., Vernon, California, Fire Department; Alexander C. Black, Collector, Fire Alarm Memorabilia, Douglas, Arizona; Fireman Paul Blum, Engine 104, Los Angeles Fire Department; Charles Calderon, Librarian, Los Angeles County Medical Association; Captain Willy Dunn, Retd., San Francisco Fire Department; Bettye Ellison, History Department, Los Angeles Public Library; Jack M. Greenfield, Manager, Public Relations, ADT Inc., Parsippany, New Jersey; Suzanne Henderson, Doheny Library, University of Southern California, Los Angeles; David L. Krzemien, Assistant Superintendent of Fire Alarm Systems, Buffalo, New York, Fire Department; Fireman-Paramedic Vincent Marzo, Engine 79, Los Angeles Fire Department; Leah J. Matuson, Technical Documentation Consultant and formerly Marketing Communications Specialist, The Gamewell Corporation, Medway, Massachusetts; Fire Chief Robert H. Noll, Yukon, Oklahoma, Fire Department; Captain William Noonan, Boston Fire Department; William Proper, Fire Protection Engineer, Jet Propulsion Laboratory, California Institute of Technology, Pasadena, California, and formerly Assistant Engineer, National Board of Fire Underwriters; Reference Desk Librarians, Glendale, California, Public Library; Steven Scher, Alarm Historian and Authority on the Fire Department of New York; James A. Spear, Collector, Fire Alarm Memorabilia, Leroy, New York and Vicky Stills, Special Collections, University of Southern California, Los Angeles.

Paul Ditzel
Woodland Hills, California

BIBLIOGRAPHY

Berst, Charles, District Sales Manager, Southern States, Gamewell Co. "The Development of the Fire Alarm Box, A Short History from the Primitive First Type to Modern Unit — Crudeness of the First Boxes — Great Advances Made". **Fire and Water Engineering,** November 19, 1924.

"Boston Fire Alarm Telegraph." **Gleason's Pictorial Drawing Room Campanion.** 1852, p. 269.

"Boston Fire Alarm Telegraph," **The Firemen's Advocate,** May 5, 1857; "The Telegraph Fire Alarm," May 9, 1857.

Brayley, Arthur Wellington. **A Complete History of the Boston Fire Department, Including The Fire-Alarm Service and The Protective Department.** Boston: John P. Dale & Co.; 1889.

"Call Startles FireFighter". **Buffalo Courier-Express,** September 6, 1955.

Carrell, John. "San Francisco's Electrical Fire Protection." **Journal of Electricity, Power and Gas.** Volume XXXVI, No. 3.

Channing, William F. "The America Fire-Alarm Telegraph." **Ninth Annual Report of the Board of Regents, Smithsonian Institution.** Washington, D. C.: Smithsonian Institution: 1855, pp. 147.

_____ . **"On The Municipal Electric Telegraph; Especially In Its Application to FIRE ALARMS."** Extracted from **The American Journal of Science And Arts."** Volume XIII, Second Series. New Haven: Yale College, 1852.

Chester, Stephen. **Fire-Alarm Telegraph, Improved and Perfected.** New York: E. O'Keefe, Printer and Stationer: 1876.

"Coded Telegraphic Fire Alarms 111 Years Old." **The Municipal South,** p. 16.

"Detroit's Early Fire Alarm System." **Detroit Firemen's Year Book.** N.D., pp 54.

Detzer, Karl. "In Case of Fire..." **The Reader's Digest,** September, 1954; pp. 41.

"Development of the Fire Alarm Signal Box." **Fire And Water Engineering,** January 14, 1914; p. 18.

Ditzel, Paul. **A Century of Service, The Centennial History of the Los Angeles Fire Department.** New Albany, Indiana: Conway Enterprises, Inc.; 1986.

_____ . "Buffalo's Bad-Luck Box." **Firehouse,** February 1983, pp. 30.

_____ . **Fire Engines, Firefighters.** New York: Crown Publishers, Inc., 1976.

_____ . **Fireboats, A Complete History of the Development of Fireboats in America.** New Albany, Indiana: Conway Enterprises, Inc.; 1989.

_____ . **Firefighting, A New Look in the Old Firehouse.** New York: Van Nostrand Reinhold Co.; 1969.

_____ . "Old 'Hoodoo' Fire Alarm Box Had Long and Tragic Record." **Fire Engineering,** June, 1949, p. 426.

_____ . "The Hoodoo Box." **The Buffalo News Magazine,** May 15, 1983, pp 19.

"Fire Alarm System." **The Grape Vine,** Los Angeles, June 30, 1934, pp 1.

"The Fire Alarm Telegraph Circuit." **Fire Engineering,** Part 1, August 1966, pp. 128; Part 2, September 1966, pp. 43.

"Early History of ADT Organization." **The ADT Transmitter,** November-December, 1968, pp. 4.

Fuszara, Walter. "'Doorman Is Sought In Blaze'." **Buffalo Evening News,** December 31, 1981, p. 1.

Galway, John, Chief Operator, Boston Fire Alarm Office. "A Brief History of the Fire Alarm Telegraph," written especially for the New England Signal Association, circa 1933. (Reprinted in the **Associated Municipal Signal Service Journal,"** circa 1933.

"Gamewell Co. Has Been Indicted. Charged With Monopolizing Fire Alarm Business." **The Journal of Commerce,** November 15, 1946.

The Gamewell Fire Alarm Telegraph Co. **The American Fire Telegraph.** Boston: Franklin Press: 1879.

_____ . **Fire Alarm Telegraphs.** Catalog, New York: 1855 - 1909.

_____ . **Emergency Signaling.** Catalog, New York: 1855 - 1916.

_____ . **Municipal Fire Alarm Systems.** Catalog, New York: 1915.

_____ . Numerous catalogs, sales literature, brochures, corporate business letter, pamphlets, and advertisements. N.D.

_____ . Twenty-Seventh Annual Report, The Gamewell Company, founded 1855, and subsidiaries, Rockwood Sprinkler Company, Eagle Signal Corporation. 1951.

Heath, Sarah Gamewell. **"Family History Compiled from Family Records from 1900 to 1910."** Updated by members of the Gamewell-Jenkins Family. (N.B.: Excellent geneological material on Gamewell family.) Unpublished manuscript in Camden, South Carolina, Archives. N.D.

Heiser, H. M. **Memories of the Fire Service.** Publisher Unknown; 1941.

Hennessy, William F. "Alarm Box Info, Knowledge of Essential Parts of Various Types Fire Boxes Necessary For All Members." **W.N.Y.F.**, July, 1946, pp. 19.

Kirkland, Thomas J. and Kennedy, Robert M. **Historic Camden, Part Two, Nineteenth Century.** Camden, South Carolina: Kershaw County Historical Society, 1965.

Konov, Patricia Jenkins, Ph.D. "Explanations of Family Journals." (N.B.: Excellent geneological material on Gamewell.) Unpublished manuscript in Camden, South Carolina, Archives. N.D.

Masterson, Joseph S., Retired Fire Commissioner, Buffalo, New York and Ditzel, Paul. "Hoodoo Box Number 29." **For Men Only,** October 1955, pp. 28.

Mulrine, Joseph F. "The Sprinkler, First Line of Defense For 257 Years." **Firehouse,** July, 1980, pp. 40.

"The Municipal Telegraph." **The Commonwealth,** December 30, 1851.

O'Banion, Albert Lee, Superintendent Fire Alarm, Boston, Massachusetts. "Boston Fire Alarm System: First — and one of Finest — in America." **Municipal Signal Engineer,** May, June, 1947.

"Official Antique Fire Alarm Telegraph Box Collector's Guide." New Albany, Indiana: Conway Enterprises, Inc: 1989.

Pearson, Ronald. "From Boxes to Bells and Gongs." **The Extra Alarmer, Newsletter of the Twin Cities,** (Minneapolis and St. Paul.) Volume 11, Nos. 2, 3, 4, 5, 6; June, 1984 - March, 1985.

_____ . "Silent Sentinel." **The Extra Alarmer, Newsletter of the Twin Cities,** Minneapolis and St. Paul.) Volume 7, Nos. 8 - 12, November 1980 - March, 1981: Volume 8, Nos. 1 - 6, April, 1981 - March, 1982; Volume 9, Nos. 1- 2, April, 1982 - June - July, 1982.

Perrin, Jan. "Call Box Recall, Telephones do the job in L.A." **Firehouse,** May, 1987, p. 37.

Raymond, F. A., Electrical Engineer, National Board of Fire Underwriters. "Signaling Systems." A Lecture Before the Insurance Institute of America. New York: January 31, 1922.

Sendler, S. Gerald, M.D. "DOCTORS AFIELD, William Francis Channing, (1820 - 1901)." **The New England Journal of Medicine,** Volume 267, No. 10, pp 501; September 6, 1962.

Sassaman, Richard. "What Hath Morse Wrought." **American History Illustrated,** April, 1988, pp. 46.

Scher, Steven. "Dates Relating to The Fire Alarm Telegraph Service in New York City." Privately published, 1986.

_____ . "The Call Box, 90 Years of Change." **Firehouse,** November, 1979, pp. 46.

Shelley, Frederick. **Aaron Dodd Crane, An American Original.** Columbia, Pennsylvania: National Association of Watch & Clock Collectors, Inc.; Summer, 1987.

Smith, Robert. "Bells Will Toll No More." **Firehouse,** December 1983, p. 67.

Supple, Jack. **History Of The Buffalo Fire Department, 1880-1979.** Privately published, 1980.

Swim, Grenfell A. "The Second Hundred Years Of Fire Alarm." **Municipal Signal Engineer.** March - April, 1954, pp. 9.

U. S. Department of Justice. "Antitrust Division criminal and civil judgments against certain fire alarm systems manufacturers and their officers:. Final Judgment and Consent Decree, District Court of The United States For the District of Massachusetts. Civil Action No. 6150. March 22, 1948.

United States Fire and Police Telegraph Co. **Complete Fire and Police Telegraph Systems.** Boston: 1897.

Warren, Daniel, editor. "History and Description of the Boston Fire Alarm Telegraph." **The Firemen's Advocate,"** December 18, 1858.

Werner, William. **History of The Boston Fire Department And Boston Fire Alarm System.** Boston: The Boston Sparks Association, Inc., 1974.

"Whatever Happened to Fire Alarm Boxes: Why They're Vanishing." **U. S. News & World Report, May 15, 1978, p. 64.**

INDEX

Note: page numbers shown in italics denote pictures or picture captions.

CI denotes color insert

Additional Book Selections From the FIRE SERVICE HISTORY SERIES

Fire Buff House Publishers
Division of Conway Enterprises, Inc.
P.O. Drawer 709 • New Albany, Indiana 47151
1-800-457-2400

"A CENTURY OF SERVICE"
THE FASCINATING STORY OF THE

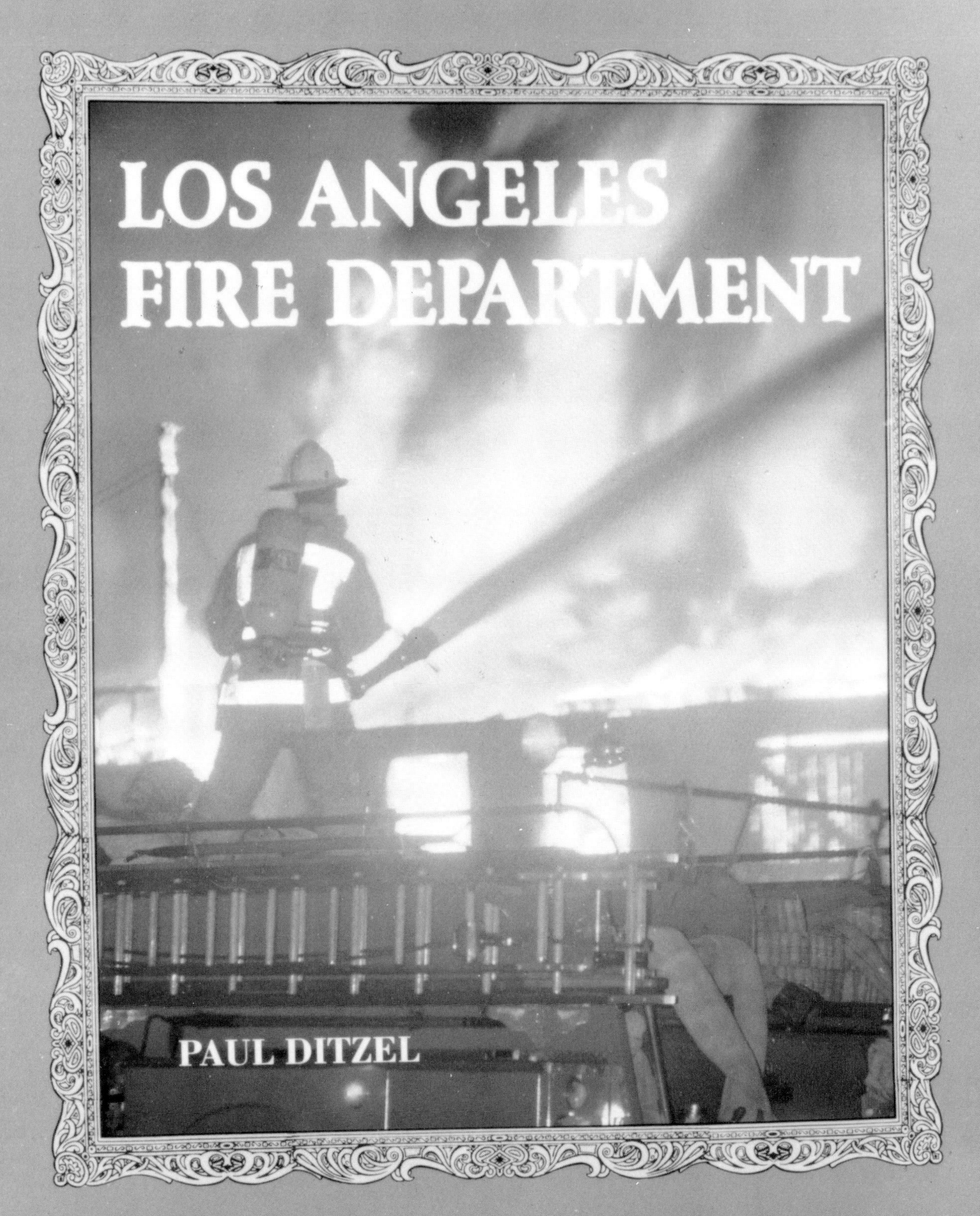

HUNDREDS OF SPECTACULAR ACTION PHOTOS
OF FIRES AND APPARATUS

Fire Service History Series
from
FIRE BUFF HOUSE

A CENTURY OF SERVICE

"A CENTURY OF SERVICE, *The Centennial History of the Los Angeles Fire Department,* is a **brilliant** historical record of one of America's great fire departments. It commences with the first day the Los Angeles paid fire department came into existence — February 1, 1886, and you are there. At least you get that feeling when the company makes its first run. The tremendous amount of research that Ditzel put into this book is evident at the very outset and it is not written in the usual form, listing just dates and events. He tells the story of the LAFD in a narrative form that ties everything together and makes you wonder...where did he get all the information? The 264 pages of this book contain hundreds of colorful photographs of fires and apparatus ...This book is not just a 'picture book'. It is a massive research project filled with fascinating facts and great commentary and rare apparatus stories. If your budget allows for the purchase of one book, then be sure that this is the one!" — Battalion Chief Bob Burns, Philadelphia Fire Department, Retd., in his Firehouse Magazine review.

"A highlight of the past year's fire service publications comes out of Los Angeles where noted fire service author and historian, Paul Ditzel, has produced an outstanding history detailing the past 100 years of the department serving America's second largest city. Originally issued as part of a centennial history yearbook by the Los Angeles Firemen's Relief Association, the original book was a near instant sellout and is already rated by book collectors as a rare volume. The good news about the Ditzel epic, however, is that the well-illustrated history section of the original volume has been reprinted and republished by Fire Buff House in New Albany, Indiana. For those interested in classic fire department history, this volume is a collector's 'Must Buy' as it tells the story in both words and pictures of the LAFD in about as accurate and concise a format as is possible. Author Ditzel...has earned his reputation as a Dean of Fire Service Historians with over 30 years of authorship of firehouse lore to his credit." — Jeff Schielke, The Fire Buff's Bookshelf, *The Visiting Fireman, 1988.*

"It is perhaps Paul Ditzel's finest work — the crowning touch of a lifetime of writing excellence." — Captain Tony DiDomenico, Los Angeles Fire Department Fireman of The Year 1985.

"A CENTURY OF SERVICE is super! I have been reading it a little at a time, savoring each detail...A masterful job of weaving everything together." — A.K. Rosenhan, Consulting Engineer, Mississippi State University.

"The first complete history of the Los Angeles Fire Department is told in this book of more than 250 pages and containing hundreds of spectacular action photos of fires and apparatus...*A Century of Service,* two years in preparation by America's foremost fire historian...is more than a history of the Los Angeles Fire Department and the Volunteer Fire Department which preceded it. The author links the development of the LAFD with historic, social, political, economic events and apparatus evolvement in other United States cities. HIGHLY RECOMMENDED." — The Fire Brigade Helmet Collector, Amsterdam, The Netherlands.

"A CENTURY OF SERVICE, *The History of the Los Angeles Fire Department,* is an extra-large format volume...authored by Paul Ditzel, one of the best-known writers in the fire service, and carefully researched...it gives a true picture of this West Coast department noted for being progressive and setting the standard for many departments throughout the United States. It contains hundreds of photographs of spectacular fires and apparatus." — The California Fireman.

FIREBOATS

Paul Ditzel

"Fireboats: A Complete History of the Development of Fireboats in America." North American Society For Oceanic History 1989 JOHN LYMAN BOOK AWARD. Honorable Mention In Marine Science and Technology.

"For 123 years, the boats and the firefighters who man them have contributed their share to the heroic legends of the American fire service. Until now, there never has been a complete account of their incredible deeds and the proud record of the nation's firefighting fleet. Paul Ditzel corrects that oversight with *Fireboats,* a superbly written and fully-illustrated history of these amazing vessels and the gutsy crews who power them into battle. Ditzel literally covers the waterfront — from ocean ports on the East, West and Gulf Coasts to the inland harbors of the Great Lakes...Fortunately, the technical explanation doesn't get in the way of the story and it's the city-by-city, boat-by-boat deeds of daring that makes this a fascinating book for any reader. The photography is outstanding, with plenty of action shots of the boats doing what they do best — fighting fires on ships, piers and in waterfront buildings. Every boat, past and present, has its own heroic stories and Ditzel tells them well...Ditzel...is widely respected as a fire historian. *Fireboats* adds luster to his well-deserved reputation." — Hal Bruno, Contributing Editor, Firehouse Magazine and ABC News Political Director.

"Paul Ditzel, who was the Los Angeles Fire Department's first civilian fire inspector, has written many books on the history of the fire service. The text of this...hardcover book is very descriptive and easy to understand, and is equal to any of his previous works. It is obvious that a tremendous amount of research was necessary to accumulate both the historical data and the 225 truly outstanding photographs and illustrations, many of them said to be previously unpublished. The numerous fires and incidents involving fireboats that Mr. Ditzel describes are certainly historic and interesting and provide information for historians and fire buffs. But beyond that, they will teach students and even experienced firefighters a great deal about a subject that is not very well known or appreciated in the fire service...The text of this book thus becomes an invaluable educational aid...The comprehensive text in this book brings out many of the essential features of waterfront firefighting and I strongly recommend it to those seeking proficiency in this aspect of the fire service. The many firefighting experiences enumerated can be used as effective standard operating procedures' for waterfront operations." — Leo D. Stapleton, Fire Commissioner/Chief of the Boston, Massachusetts, Fire Department writing in the Fire Journal.

"This book does more than document facts and figures — it makes fireboats come alive, as the heroic and dangerous duties of firefighting mariners fill the pages. Each fireboat has its own special and unique history. This book is for them and the firefighters who race to burning ships and blazing piers with the spray flying over the decks and the monitors blasting." — James P. Delgado, Chief Maritime Historian, National Park Service and Head of the National Maritime Initiative.

"An excellent job...I believe you are the most accurate writer I have shared information with." — Fire Chief Stanley L. Thaut, City of Tacoma, Washington, Fire Department.

"I was impressed with the vast amount of information and the photographic material." — Battalion Chief William J. Guido, Marine Division, Fire Department of New York.

"As usual, Paul Ditzel has outdone himself. *Fireboats,* like all his books, is well-researched and wonderfully written. It is indeed an Eye-Opener. We cannot recommend this book highly enough. It is Superb!" — Jack Robrecht, Philadelphia Fire Department Historian, The National Firehouse and Museum of Philadelphia.

"Paul Ditzel is not only the preeminent authority on the fire service, but the most vivid and engrossing writer on the subject. I always look forward to his publications." — Jerome F. Brown, Ph.D., Department of History, New Mexico State University.

"The name of the author says it all. As the Dean of Fire Service Authors, most everyone is familiar with the quality of his research and writing. Whether or not you have ever seen a fireboat, this book will hold your interest and be a valuable addition to your library." — A.K. Rosenhan, Consulting Engineer, Mississippi State University, writing in Fire Chief Magazine.

"Naval and other maritime historians will find Ditzel's 'Log of American Fireboats," included as an Appendix, to be of deep historical interest...Never before has similar material been made available to historians, naval and marine architects, boating and firefighting buffs, as well as model builders. Ditzel, author of more than 13 other books — many of them on fire history — makes the reader feel he is on-deck and in the thick of the smoke and heat as fireboats do battle against flaming waterfront disasters...With its sometimes technical approach to the subject, the book belongs on the shelves of libraries and universities, as well as individuals interested in the history and science of fireboat design and construction." — Oklahoma State University, Headquarters for the International Fire Service Training Association.

"If you're fascinated with American fireboats, this is **the** book for you. Hailed as "A Complete History of the Development of Fireboats in America," this book contains detailed and technical text plus photos, diagrams and marine blueprints." — The Westcoast Mariner.

"I have just finished reading...*FIREBOATS*. Outstanding is the only word that fits the bill...Research was tremendous...a staggering volume of vessel data." — Alex. C. Meakin, Chairman, Board of Trustees, The Great Lakes Historical Society.

"This big picture-textbook is the first volume ever to trace the history of fireboats in America. Author Ditzel knows his ships and his men and his history...Told in a gripping manner both visually and narratively." — The Book Reader.

"Paul Ditzel is well-qualified to write about fireboats as he is a prize-winning author and historian...If you really like boats, this is a must for your library." — The Tug's Wake, published by the International Retired Tug Association.

"Paul Ditzel, a well-known writer on fire service subjects, has done it again...Almost everything you've always wanted to know about fireboats and didn't know who to ask is answered in this book...With the cost of most books today, this book is a bargain and a must for any fire buff's library." — Curt Elie, The Wagon Pipe Newsletter of the Friendship Fire Association, Washington, D.C.

"This book is one of the best researched books that I have read. The author did an excellent job. It is a book that anyone with an interest in marine articles should have in their collection. I have read the book twice...The photos that are included...are great and all the history just cannot help but to hold your interest." — Bob Martin, Pacific Coast Tug Society.

"It is a splendid book and will make a fine addition to our collection." — Mary Blackford, Librarian, Maritime Museum of the Atlantic, Halifax, Nova Scotia.

"*FIREBOATS* neatly fills the gap that has long existed in the documentation of this particular slice of Americana and fire service history. With the holidays just around the corner, now is the time to treat yourself or your favorite firefighter, fire buff, or mariner with a new addition to the bookshelf that will long be enjoyed." — Vincent Marzo, The Firemen's Grapevine.

NORTH AMERICAN SOCIETY FOR OCEANIC HISTORY

1989

John Lyman Book Award

HONORABLE MENTION

IN

MARINE SCIENCE AND TECHNOLOGY

PRESENTED TO

PAUL DITZEL

FOR

FIREBOATS: A COMPLETE HISTORY OF THE DEVELOPMENT OF FIREBOATS IN AMERICA

President

Chairman,
Awards Committee

Dr. John Lyman, 1915-1977, a founder of NASOH, described himself as a Consultant on Maritime History, Nautical Vexillology, and the Ocean Environment. He wrote extensively for maritime journals and for many years published Log Chips, *which recorded the histories of ships and shipyards.*

FOR HALF A CENTURY - 1872 TO 1922 - CHEMICALS EXTINGUISHED 80% OF OUR FIRES

CHEMICAL FIRE ENGINES

W. FRED CONWAY

MORE THAN 100 OLD PHOTOGRAPHS AND DRAWINGS

Fire Service History Series from FIRE BUFF HOUSE

FIREFIGHTER'S News

CHEMICAL FIRE ENGINES, by W. Fred Conway

If ever a labor of true love were to need defining, I would commend the person in search of the definition to contact the author of this fine reference text.

Mr. W. Fred Conway, a former volunteer fire chief and fire safety products purveyor, has obviously done a great deal of homework in bringing this book to the market. The subject of his affections is but one small segment of the history of the American Fire Service, the chemical fire engine. However, Mr. Conway's depth of knowledge, liberal use of period photos and drawings, not to mention a wealth of original manufacturer's specifications, have come together in the fine fashion.

The reader is transported back in history to 1872, a time when hand-drawn equipment formed the backbone of our operating scheme. Mr. Conway then brings you from that period to the end of the chemical engine era more than 50 years later. All of the major manufacturers and most of the others are given space in this book.

I was particularly impressed with the quality of Mr. Conway's reproduction of period literature. Usually such material suffers from the transfer to modern print. This is not the case here. Complete instructions for use of many of the chemical engines gave me a view of the distant past that brought a warm glow to my cynical modern psyche. In my mind's eye, I was a Captain in the fire department of 1895; controlling potential disasters with a few gallons of water, powered by carbon dioxide generated when the mixture of water, sodium carbonate and sulfuric acid came together in my 40-gallon copper tank.

This book will make a welcome addition to the book shelf of every serious fire science student and fire buff alike.

Chemical Fire Engines by SPAAMFAA member W. Fred Conway fills a significant gap in antique fire apparatus literature. During the transition from volunteer companies and their hand pumps to paid fire departments with steam fire engines, a need developed for an apparatus which could handle the small fire quickly. The steam fire engine required several minutes to build up the steam pressure needed to power the pump and produce a good stream, and the stream produced was much larger than required for many fires. The chemical engine with capacities of 25 to 200 gallons was the solution to the small fire problem. It also took care of the situation faced by the small town volunteer department: i.e., how to provide reasonable fire protection where the number and magnitude of fires would not justify the capital cost and continuing operating cost of a steam fire engine. (Horses had to be fed every day whether there was a fire or not .)

Chemical Fire Engines provides a good insight into the evolution of the chemical apparatus and the unsubstantiated hype that accompanied its promotion."Chemical has 40 times the fire fighting efficiency of water". Testimonials from fire Chiefs with Champion or Babcock chemical Engines in their departments, which were solicited by the Fire Extinguisher Manufacturing Company of Chicago in 1895, showered praise on the chemicals and make one wonder why any other apparatus was ever needed. Many black-and-white pictures of hand drawn, horse drawn, and motorized chemicals show most of the different types. Much of the information is in the form of manufacturers' literature which provide specifications and pictures. Pirsch, American LaFrance, and Obenchain-Boyer are among the familiar manufacturers whose products are shown.

The demise of the chemical started in 1913 when Charles H. Fox developed a booster system composed of a small centrifugal pump mounted in front of the radiator for the gasoline engine. Water was supplied by a water tank behind the driver's seat. Although Ahrens-Fox delivered ten booster equipped fire engines to the Cincinnati Fire Department in 1913, the chemical tank continued to be used into the 1920's with the last known chemical engine being manufactured in 1934.

Fred Conway's interest in and research on the chemical engine has provided a very enlightening book on a segment of antique fire apparatus, usually limited to an occasional picture in fire department histories.

American Fire Journal

Buff Book Takes You Back to the Time of Chemical Engines

A big seller with nostalgia buffs over the years has been a reprint of a turn-of-the-century Sears & Roebuck catalog, with its anachronistic wares, heavy gothic illustrations and lettering and pre-inflation prices. Fire nostalgia buffs will get a similar kick out of **Chemical Fire Engines** by **W. Fred Conway,** former fire chief, founder of Home Safety Equipment Co., Inc. and himself the owner of three restored chemical fire engines.

The book takes a loving look at the pressurized - carbon - dioxide - carrying rigs in use from 1872 to the 1930s. Every facet of their development, acceptance and use is detailed by advertisments, descriptions, excerpts from actual users' surveys and, of course, lots of illustrations.

What sets Chemical Fire Engines apart from other fire history books is that it doesn't read like a history report written by a modern author, but instead has the look and feel (complete with old-fashioned type and actual turn-of-the-century-style drawings and photos) of a book written in the engines' heyday.

W. Fred Conway is obviously an avid afficionado of chemical fire engines, and his interest and authority come through in the book. Whatever you want to know about this fascinating niche in fire history, it is here.

FIRE APPARATUS JOURNAL

For more than half a century, chemical fire engines extinguished nearly 80% of all fires in cities, towns and villages across America. Yet little has been written about these interesting apparatus until now. W. Fred Conway has thoroughly researched and documented the story of the chemical engine in this new book appropriately titled "Chemical Fire Engines". The book is hardbound with over 125 pages. Throughout this volume there are photos, illustrations and reproductions of period manufacturers' literature which provides a wealth of knowledge on these apparatus. From the earliest chemical engines in the 1870's through "modern" motorized rigs of the 1920's, every conceivable type and make of chemical rig is covered in depth. This book has a definite place in any apparatus historian's library as it traces the development of a particular type of fire apparatus that served a much needed function until the development of booster hose and equipment on later pumpers.

FIRE SERVICE DIGEST

"Chemical Fire Engines" by W. Fred Conway is a 128 page book that answers a lot of questions and provides plenty of information and illustrations on one of the least-known aspects of fire service history. I was well-pleased by the details and reprints of old advertisements, catalogs, etc. This is full of well-researched unusual facts and stories, some of which surprised me. I highly recommend this hardcover book.